Human Factors in Alarm Design

Human Factors in Alarm Design

Edited by

Neville Stanton

University of Southampton

Taylor & Francis
Publishers since 1798

UK	Taylor & Francis Ltd, 4 John St, London WC1N 2ET
USA	Taylor & Francis Inc., 1900 Frost Road, Suite 101, Bristol, PA 19007

British Library Cataloguing in Publication Data
A catalogue record for this book is available from the British Library

ISBN 0–74840–0109–1

Library of Congress Cataloging in Publication Data are available

Cover design by Amanda Barragry

Set in 10/12pt Times
by Graphicraft Typesetters Ltd., Hong Kong

Printed in Great Britain by Burgess Science Press, Basingstoke on paper which has a specified pH value on final paper manufacture of not less than 7.5 and is therefore 'acid free'.

Contents

Preface

I believe that alarm displays deserve special consideration in design. This is because of their intended usage: i.e. to communicate information that is of crucial importance for overall system integrity. However, alarm systems may not always achieve this objective. This book sets out to present some of the current Human Factors (or Ergonomics if you prefer) thinking on alarm design. As the examples given in the text illustrate, there is clearly a need for input from ergonomists and human factors engineers. The main aim of this book is to show how this might be done and indicate the contribution to be made in a variety of settings. Although most of the work presented concentrates on alarm systems in process industries, other areas presented include: aviation, automobiles and intensive care. I believe that there is much to be gained from the transfer of knowledge between the work conducted in these different application areas.

The contents of this book are based upon the contributions to a one-day conference on alarm design held at Aston University on Friday 2 October 1992. This conference was organized by myself on behalf of The Ergonomics Society. Inspiration for the book came from my research into alarm systems over the past five years. During this time, I became aware that, despite considerable work conducted in this area, no-one had yet attempted to bring the work on alarms together into a single volume. Following the success of the conference and the subsequent requests for reprints of the proceedings, I decided to produce this book. The book has also benefited from one additional chapter produced subsequent to the conference. This is from David Woods who agreed to write a chapter on reasoning in dynamic fault management.

The book is structured into four main sections: experimental research into alarm design, considerations of the human operator, design and evaluation of alarm systems, and applications of alarm systems. Each section contains three chapters. The book begins with an introductory chapter and ends with a concluding chapter. For the reader there are different ways they may choose to approach this book. It can be skimmed at a high level by just reading the section introductions. This understanding can be broadened by reading the introductory and concluding chapters. Greater depth can be added by reading individual chapters. Although there is some cross-referencing, each of the chapters can be read as a stand-alone source. However, a fuller appreciation of this topic can only be gained by reading a substantive part of this book.

I hope that the reader will appreciate the balance between research and practice that this text attempts to achieve. Given the applied nature of the work, it is hoped that the book will be of interest and use to academic researchers and industrial practitioners alike. This balance is reflected in the contributors, i.e. half from industry and half from academia. I also hope that the reader will leave this text as enthusiastic about the contribution that human factors (cf. ergonomics) has to make in alarm design as I am.

I would also like to thank those individuals who have made this book possible: the individual contributors (in order of appearance: Judy Edworthy, Paul Hollywell, Ed Marshall, Tom Hoyes, Andreas Bye, Øivind Berg, Fridtjov Øwre, Harm Zwaga, Hettie Hoonhout, David Usher, David Woods, Ned Hickling, Sue Baker, Chris Baber and Tina Meredith), the staff of Taylor & Francis (Richard Steele, Wendy Mould and Carolyn Pickard), The Ergonomics Society (specifically Jonathan Sherlock, Dave O'Neil and David Girdler), Sue Davies (for patiently typing my contributions) and Maggie Stanton (for her support and proof reading). Finally, with a note of sadness, I would like to dedicate this book to the memory of my friend and colleague Dr Tom Hoyes (1964–1993) who was tragically killed in a car accident earlier this year.

Neville Stanton
August, 1993

Contributors

Dr Chris Baber
Industrial Ergonomics Group
School of Manufacturing and
Mechanical Engineering
University of Birmingham
Birmingham, B15 2TT

Dr Sue Baker
Safety Regulation Group
Civil Aviation Authority
Psychology Division
RAF Institute of Aviation
Medicine
Farnborough
GU14 6SZ

Dr Øivind Berg
Institutt for Energiteknikk
OECD Halden Reactor Project
PO Box 173
N-1751 Halden
Norway

Dr Andreas Bye
Institutt for Energiteknikk
OECD Halden Reactor Project
PO Box 173
N-1751 Halden
Norway

Dr Judy Edworthy
Department of Psychology,
University of Plymouth
Drake Circus
Plymouth, Devon
PL4 8AA

Mr Ned M. Hickling
PWR Projects Group
Nuclear Electric
Booths Hall
Knutsford, WA16 8QG

Mr Paul D. Hollywell
EWI Engineers and Consultants
Electrowatt House
North Street
Horsham, West Sussex,
RH12 1RF

Dr Hennie C.M. Hoonhout
Utrecht University
Psychological Laboratory –
Ergonomics Group
Heidelberglaan 2,
3584 CS Utrecht
The Netherlands

Dr Thomas W. Hoyes (deceased)
Human Factors Research Unit
Aston University
Birmingham, B4 7ET

Mr Edward Marshall
Synergy, 14 Northcote Close,
West Horsley, Leatherhead,
Surrey, KT24 6LU

Ms Christina Meredith
University of Plymouth
Drake Circus
Plymouth, Devon
PL4 8AA

Dr Fridtjov Øwre
Institutt for Energiteknikk
OECD Halden Reactor Project
PO Box 173
N-1751 Halden
Norway

Dr Neville Stanton
Department of Psychology
University of Southampton
Highfield
Southampton, SO9 5NH

Dr David M. Usher
InterAction
19 Northampton Street
Bath, BA1 2SN

Dr David D. Woods
Cognitive Systems Engineering
Laboratory
The Ohio State University
Columbus, OH43210, USA

Dr Harm J.G. Zwaga
Utrecht University
Heidelberglaan 2
3584 CS Utrecht
The Netherlands

1

A human factors approach

Neville Stanton

The need for research

This book addresses the human factors concerns of alarm systems. Lees (1974) noted that the design of industrial alarm systems was an area worthy of research when he wrote:

> Alarm systems are often one of the least satisfactory aspects of process control system design. There are a number of reasons for this, including lack of a clear design philosophy, confusion between alarms and statuses, use of too many alarms, etc. Yet with the relative growth in the monitoring function of the operator, and indeed of the control system, the alarm system becomes increasingly important. This is therefore another field in which there is much scope for work.

The need for basic research into alarm system design has been made even more necessary by recent legislative requirements. For example EC Directive 89/391 which covers alarm systems under the umbrella of 'work equipment used by workers'. The directive states that:

> Warning devices on work equipment must be unambiguous and easily understood.

These points give purpose to this book. The main tenet of the book is that industrial alarm systems have severe shortcomings in human factors terms, i.e. they are ambiguous, they are not easily perceived, nor are they easily understood. These are all issues where human factors can, and should, make a significant contribution.

What is an alarm?

There is a need to develop an accurate definition of the term 'alarm', because unless the subject under analysis is clearly pinpointed it cannot be studied

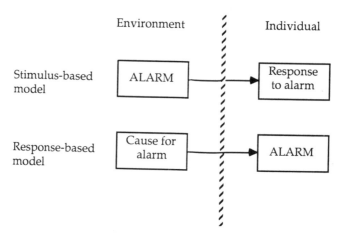

Figure 1.1 'Simple' definitions of alarm.

properly. This is done by first considering previous definitions and noting what is wrong with them. The term 'alarm' is to be found in daily use in many forms.

The common usage of the term may give the impression that its use is well understood. However, further consideration suggests that it is not so clear cut. A frequently given definition of an alarm is 'a significant attractor of attention', however a dictionary (Collins, 1986) gives nine definitions of the word 'alarm'. These are:

- to fill with apprehension, anxiety or fear;
- to warn about danger: alert;
- fear or terror aroused by awareness of danger: fright;
- a noise, signal, etc., warning of danger;
- any device that transmits such a warning: a burglar alarm;
- the device in the alarm clock that triggers off the bell or buzzer;
- a call to arms;
- a warning or challenge.

The above definitions demonstrate the inadequacy of the first definition, because whilst an alarm may attract attention, its 'attractiveness' is only one of its many possible properties or qualities. Therefore the main problem with definitions of the term 'alarm' is that they tend to concentrate on only one or a restricted range of the qualities. Thus there is the need to consider the term further, to unpack and understand the nature of an 'alarm'.

Figure 1.1 indicates why there is a problem in defining an alarm. The term may be used to define both the stimulus and the response on different occasions. In the stimulus-based model an alarm exists in the environment and its presence has some effect on the individual, whereas in the response-based model, the stimulus causes an alarm state in the individual. The first model

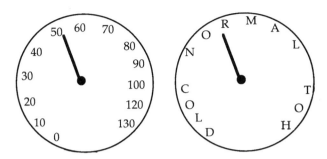

Figure 1.2   An example of quantitative (left) and qualitative (right) displays.

suggests that alarms are relatively homogeneous: they can be clearly identified by all; whereas the second model suggests that different individuals may find different stimuli 'alarming'. Therefore there may be disagreement between individuals over what constitutes an alarm situation, based on their experiences of it. The stimulus-based model characterizes the engineering approach, i.e. the assumption that the alarm will mean the same thing to all people, whereas the response-based model characterizes the psychological approach, i.e. people interpret situations differently, and that their reaction will be based upon this interpretation.

Brief historical perspective

The notion of an alarm has been around since the dawn of mankind. 'Alarms' may be viewed as fundamental to the fight–flight principle; the alarm prompting a state of arousal that requires the human to respond in an appropriate manner, either to run from the attacker or to stay and fight for life. Alarms or warnings have existed in the form of cries for help when an individual is attacked, ringing of bells to inform people that a town is under siege, and the ringing of a handbell by a town crier prior to presentation of important information. Since the industrial revolution technology has introduced new kinds of problems for mankind. There has become the need to inform on things that are not directly visible to the naked eye, such as steam pressure, oil temperature, etc. This information was typically presented via dials. The type of display can provide quantitative or qualitative readings (Oborne, 1982) (Figure 1.2). For example, temperature can be presented as degrees Celsius, requiring someone to read the value and interpret it as too cold, normal or too hot. A qualitative display may simplify this task by presenting bands on the dial which are marked, 'cold', 'normal' and 'hot'. Then all that person has to do is observe within which band the needle lies. This type of display also provides trend data, i.e. the observer can watch the relative position of the needle throughout operating conditions. However, it was soon noticed that the useful information was binary in nature, i.e. either everything

Figure 1.3 Oil pressure dial from steam engine.

was OK or it was not. Therefore most of the information on the analogue dial appears to be redundant. This led to the development of binary dials. Figure 1.3 shows a photograph of a binary dial taken from a steam engine built at the beginning of this century. It informs the driver on the status of the oil to valve and pistons. It is also interesting to note that the dial contains instructions to the driver on how to maintain the engine status under certain operating conditions. The legend reads:

WHEN RUNNING WITH STEAM SHUT OFF MOVE REGULATOR FROM FULL SHUT POSITION UNTIL POINTER SHOWS IN WHITE SECTION.

Clearly under certain operating conditions, the warning dial is useful to maintain correct running of the engine as it provides feedback to the driver on the state of the engine. This is in addition to its use as a warning device. It is also worthwhile pointing out that when the engine is shut down (as was the case when this photograph was taken) the dial is in its 'alarm' state, but the needle position can be simply explained by pointing to the context. Thus the nature of the information is highly context dependent. This will be a recurrent theme throughout this book.

Alarms and warning take many forms and they may have different meanings attached to them. For example, Table 1.1 illustrates some possible categories to which warnings may belong. These everyday examples of alarms and warnings suggest that 'attraction' is one possible quality of an alarm. It may attract attention but it does a lot more also. For example they can call for help, indicate that an event has occurred, call for action and communicate information. There are problems, however, as many of the alarms and warnings can be ambiguous. For example the flashing of headlights can mean 'get out of my way', 'there's a police speed trap ahead' or to indicate courtesy.

Table 1.1 Forty-nine everyday examples of alarms and warnings

WARNING:	ACTION:	INFORMATION:
Low oil light	Traffic lights	Written warnings
Low petrol light	Factory hooter	Caution
Brake lights	Gong	Turn lights
Traffic signals	Red alert	Mind-the-gap
Fog horn	Lights on	Radio paper
Lighthouse	Railway crossing	Telephone tree
Red flag	Egg timer	Hazard sign
Reversing beep	Curfew	
		HELP:
SIGNAL:	EVENT:	SOS
Police siren	Alarm clock	Whistle
Hazard lights	Burglar alarm	Hospital bleeper
Ambulance	Car alarm	Flare
Fire engine	Shoplifting alarm	999
Fork-lift truck	Bulb failure	Shout
MONITORING:	VISIBILITY:	MULTIPLE:
Baby alarm	Fog lights	Horn
Cot death alarm	Beacon	Flashed lights
Tagged criminal		

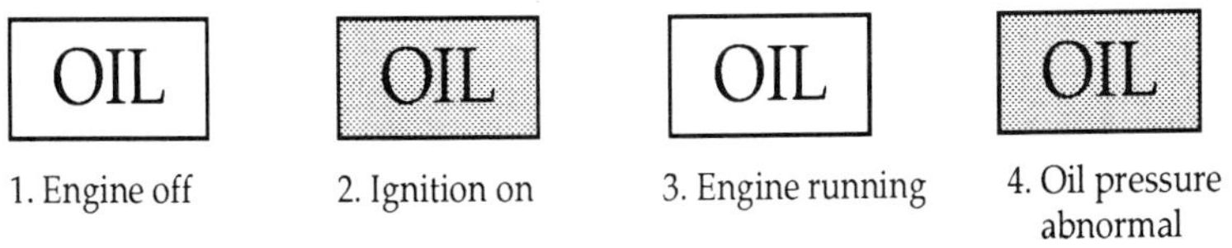

1. Engine off   2. Ignition on   3. Engine running   4. Oil pressure abnormal

Figure 1.4   States of an oil annunciator from a car dashboard.

The context of the warning can be a clue to the meaning, but there is the potential for misinterpretation. If the signal is misinterpreted it could lead to an accident.

Before developing this argument further, it is necessary to consider the context relative to the meaning of an alarm. Most readers will be familiar with in-car annunciator systems. Typically a panel of between four and twelve (normally eight) annunciators is sited in the dashboard and may be viewed through the steering wheel. The annunciator can be in any of four possible states as illustrated in Figure 1.4. These are: 1) unlit: engine off, 2) lit: ignition on, 3) unlit: engine running normally and 4) lit: oil pressure abnormal. Only in the last of these states is the annunciator in 'alarm' mode. In states 2 and 3 the annunciator is providing confirmatory evidence to the driver. In state 2 the annunciator confirms that the bulb is operational, and in state 3 the annunciator confirms that the oil pressure is normal by extinguishing the light. This represents a Boolean logic display, i.e. the state is either true or false, which is represented by the annunciator being lit or unlit in different system modes. However, unlike the dial on the steam engine, there is no analogue information such as rate and direction of change. Thus this kind of display may deprive the observer of some potentially useful information.

However, a number of problems associated with alarms have not escaped the attention of popular comedy fiction writers who parody the major inconsistencies. For example:

> ... the London night was, as usual, filled with the ringing and wailing of various alarms. . . . In fact the Detective Constable was the only person to pay any attention to the alarm bells (except of course the thousands of people with pillows wrapped round their heads screaming 'Turn them off! Please turn them off!' into the darkness of their bedrooms). *Everyone always ignores alarm bells, which is a shame,* . . . .

Elton (1991)

> Framlingham (n): A kind of burglar alarm in common usage. It is cunningly designed so that it can ring at full volume in the street without apparently disturbing anyone. Other types of framlinghams are burglar alarms fitted to business premises in residential areas, which go off as a matter of regular routine at 5.31 p.m. on a Friday evening and do not get turned off til 9.20 a.m. on Monday morning.

Adams and Lloyd (1990)

This illustrates that there is a danger that if the alarm is presented too often with no consequence, there is a tendency for it to become ignored on subsequent occasions. This is commonly known as the 'cry wolf' syndrome. The examples also raise the question of whose responsibility it is to respond to the alarm. Attending to the alarm could have negative payoffs. If it is not genuine, then the person who attends to it has wasted time and resources.

A systems model of alarms

'Alarms' can be seen to refer to various points in the flow of information between plant and user. It is generally considered that the role of the alarm is to give warning of impending danger, albeit in varying degrees of severity. Some of the definitions are shown in terms of their points along the information flow in Figure 1.5: the systems model. For example an alarm is:

- an unexpected change in system state;
- a means of signalling state changes;
- a means of attracting attention;
- a means of arousing someone;
- a change in the operator's mental state.

In Figure 1.5, transition of alarm information is shown by the arrows. If a change has occurred the operator needs to be informed about it. For example, a measured value may be beyond the limits of system threshold values, being either too high or too low. This information is sent to some means of communicating with the human operator, such as bells, flashing lights, etc.

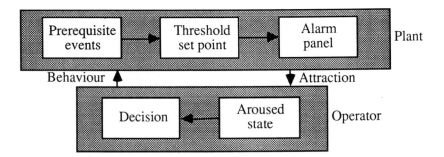

Figure 1.5   A systems model of alarms.

The operator's attention must first be drawn to the area that the alarm lies within, then the alarm has to communicate information about the event (or non-event). Based on this information the operator is required to: acknowledge the alarm (confirming that it has drawn attention) and decide what action (if any) is required, based on the information given. This may affect any subsequent operator input to the system. The systems model shows the cycle of activities that may occur between the human and the alarm. If successful, the appropriate action will be taken. If unsuccessful, then the component in the system may trip, or at extremes the whole system may shut down.

What is 'human factors'?

It has been claimed that the idea of human factors is as old as humanity, based as it is on the underlying premise that things are designed for people. Before mass production, tools would have been built for the individual user. Yet human factors is often not considered by designers and engineers (Meister, 1989). Human factors (HF) is a term that can have many meanings associated with it. In the Health and Safety Executive booklet on 'Human Factors in Industrial Safety' the term 'human factors' is defined as follows:

> The term 'human factors' is used here to cover a range of issues. These include the perceptual, mental and physical capabilities of people and the interactions of individuals with their job and working environments, the influence of equipment and system design on human performance, and above all, the organisational characteristics which influence safety-related behaviour at work.
>
> HSE (1989)

This is a very broad definition of HF, hinting at its multi-disciplinary nature. The HSE document emphasizes the need to consider the interaction between the individual, the job and the organization. This is perhaps what best characterizes human factors. Often the terms 'human factors' and 'ergonomics' are used interchangeably. Hendrick (1991) offers four main areas that ergonomics addresses to the design of human system interface technology: hardware ergonomics, environmental ergonomics, software ergonomics and macro-ergonomics. Hardware ergonomics is concerned with human physical and perceptual characteristics. Environmental ergonomics relates to human capabilities and limitations with respect to the demands imposed by the environment. Software ergonomics looks at how people conceptualize and process information. It is also referred to as cognitive ergonomics. Macro-ergonomics refers to the overall structure of the work system as it interfaces with the technology. This latter approach is in contrast to the first three in that it is 'macro' in its focus, whereas the others are concerned with 'micro-systems'.

Recent discussions of the nature of HF, have revealed that there exists some controversy over its status in research and development. Dowell and Long (1989) offer a useful tripartite classification of approaches to HF: as a craft, applied science or engineering.

- As a craft it evaluates design by comparison with previous design. Practitioners apply their previous experience in the form of rough 'rules-of-thumb'. This obviously represents a highly-skilled, but largely unstructured approach (both in terms of information content and methodology).

- As an applied science it draws on research from a number of interrelated subject areas, from psychology and physiology to computer science and engineering. It is concerned with the design of systems which can enhance human performance.

- As an engineering discipline it seeks to develop adequate design specifications and focuses on cost: benefit analysis.

These three approaches represent different views of the topic. This definition implies that research in any discipline can be craft oriented, or engineering oriented or an applied science. A craft orientation suggests that machines will be developed largely on the basis of experience of designers with previous similar machines, and rules of thumb which appear to have worked in the past. There is no guarantee that the designers' 'common sense' view of the world corresponds to that of the end user. Indeed, it is likely that someone who has had experience of the machine throughout its development cycle, i.e. a designer, will have a far more detailed view of the machine than someone who has only just met it, i.e. the user. This means that the craft approach suffers from a number of severe limitations. At the other extreme, an applied science approach could be exemplified by HF. Knowledge concerning human physical and mental characteristics could be collected empirically and applied to the design of specific machines. While such approach could produce usable machines, if conducted efficiently, it may also be costly. The engineering approach seeks to take knowledge and relate it to machine designs, so that it is possible to develop specifications. Indeed, rather than looking for generalized rules of behaviour, an engineering approach seeks to tackle specific problems. Thus, an engineering approach will be solution rather than theory oriented. The solution-oriented approach aims to propose alternatives and select the most attractive measure. However, we cannot assume that the alternatives selected are exhaustive, or that the selected measure is more than an arbitrary decision. Thus the engineering approach is quite different from the applied-science approach which attempts first to define the problem before solutions are presented.

HF is characterized by the attempt to bridge the gap between theory and application. It is relatively easy to make recommendations for improvement in design of specific tools from observing their use. However, from specific tools to other tools or systems requires a basic theory of human activity in the context of the work environment. Therefore the HF discipline will consist of:

- theories and models of human functioning;
- methods of evaluating human–machine interaction;
- techniques and principles for the application of a HF methodology.

These three points will form the basis of the rest of this chapter, and be integrated into a HF approach. This approach has been developed from individuals' experience in the field, but there are other ways of considering the discipline. The perspective chosen will obviously depend on an individual's knowledge and the information they require from the study. In addition to the perspectives provided by Dowell and Long (1989), it is possible to suggest the following four definitions of HF:

- a discipline which seeks to apply natural laws of human behaviour to the design of workplaces and equipment;
- a multi-disciplinary approach to issues surrounding people at work;
- a discipline that seeks to maximize safety, efficiency and comfort by shaping the workplace or machine to the physical and psychological capabilities of the operator;
- a concept, a way of looking at the world and thinking about how people work and how they cope.

Each of these views offers a subtly different perspective. The first suggests that 'natural laws' of human behaviour exist, which can be applied to the design and evaluation of products and environments. Whilst such a view may produce important findings it is dubious that such findings constitute immutable laws. This leads to the second viewpoint which draws on a pot-pourri of different subject matter. Alternatively, the third opinion emphasizes the need to design the job to fit the person. Problems with this approach arise from attempting to define the 'average person'. Finally, the fourth definition develops a notion of HF as an attitude: first it is necessary to recognize the need, then it is necessary to employ a body of knowledge and a set of skills to satisfy this need. The final view is distinctly different from the first three in that it proposes HF as a philosophy rather than an 'add-on' approach to design; it provides an overview of the complete design problem, rather than a discrete stage of the process.

Structure of the book

This book consists of 14 chapters in four main parts. Firstly there is an introductory chapter (chapter 1). Part one considers 'experimental research in alarm design' and has chapters on 'urgency mapping' (chapter 2), 'alarm lists' (chapter 3) and 'risk homeostasis theory' (chapter 4). Part two covers 'considerations of the human operator' and has chapters on 'dynamic fault management' (chapter 5), 'alarm initiated activities' (chapter 6) and 'supervisory control behaviour' (chapter 7). Part three looks at 'design and evaluation' and has chapters on 'an alarm matrix' (chapter 8), 'operator support systems' (chapter 9) and 'ergonomics and engineering' (chapter 10). Part four reviews 'applications of alarm systems' and has chapters on 'control rooms' (chapter 11), 'in-car warnings' (chapter 12) and 'intensive therapy units' (chapter 13). Finally the conclusions are presented in chapter 14.

References

Adams, D. and Lloyd, J., 1990, *The Deeper Meaning of Liff: a dictionary of things that there aren't any words for yet*, London: Pan.
Collins English Dictionary, 1986, 2nd Edn, Glasgow: Collins.
Dowell, J. and Long, J., 1989, Towards a conception for an engineering discipline of human factors, *Ergonomics*, **32** (11), 1513–35.
Elton, B., 1991, *Gridlock*, London: Macdonald.
Health and Safety Executive, 1989, *Human Factors in Industrial Safety*, London: HMSO.
Hendrick, H.W., 1991, Ergonomics in organizational design and management, *Ergonomics*, **34** (6), 743–56.
Lee, F.P., 1974, Research on the process operator, in Edwards, E. and Lees, F.P. (Eds) *The Human Operator in Process Control*, London: Taylor & Francis.
Meister, D., 1989, *Conceptual Aspects on Human Factors*, Baltimore: Johns Hopkins University Press.
Oborne, D., 1982, *Ergonomics at Work*, Chichester: Wiley.

Part 1
Experimental research into alarm design

Experimental research into alarm design

Neville Stanton

This section presents three chapters based on laboratory studies investigating alarm media. Chapter 2 (by Judy Edworthy) considers the construction of auditory warnings. Judy presents an auditory warning that is constructed from bursts of sound that can be repeated at varying intervals in relation to urgency of required action and the level of background noise. A burst is a set of pulses which give a syncopated rhythm – a melody that can be used to identify the nature and urgency of the warning. Pulses are tones, whose spectral and temporal characteristics can be matched to the noise environment. This type of construction is proposed as a means of developing an 'ergonomic' warning sound, i.e. one that is informative but does not unduly startle the operator. This forms the basic building block of a warning sound. From the initial burst three forms of burst are constructed. A complete warning sound is then constructed from the three burst forms. Judy presents research has been conducted into the factors of the sound construction that affect perceived urgency. For example, she asked subjects to rank sounds in terms of urgency and compared various sounds by altering 10 characteristics. The results suggest some of the characteristics of a sound that may be altered to increase perceived urgency. This work has been extended to suggest that some parameters have greater and more consistent effects on perceived urgency than others. By identifying these, urgency could be more appropriately mapped onto warning sounds that require more immediate attention and subtracted from sounds that do not require an immediate response.

Chapter 3 (by Paul Hollywell and Ed Marshall) presents an experimental study based upon an alarm handling task in a simulated control room. Control room operators typically use large arrays of visual information when controlling power station plant. Alarms are an integral part of the display system to alert the operator of abnormal plant conditions. Information technology has enabled this information to be presented in new ways, e.g. as text messages presented via a visual display unit (VDU). The rapid onset of a large number of alarm messages during a major plant disturbance can impede the operator in performing a rapid and accurate diagnosis. Therefore it was considered important by Paul and Ed to determine the rate at which operators are able to assimilate alarm information from text messages presented via a VDU in order to provide guidelines for alarm system design. In the study presented, power plant operators were observed reading and categorizing alarm messages. The results indicated that the maximum rate at which operators can read alarm messages was in the order of 30 messages per minute, whilst their preferred rate was approximately 15 messages per minute. Degradation in performance was manifested in the number of missed messages, which rapidly increased as the alarm presentation rate exceeded maximum response capacity. Paul and Ed warn of the potential mismatch between alarm presentation rates and reading performance. They propose that design consideration should include alarm presentation rates.

Chapter 4 (by Tom Hoyes and Neville Stanton) examines risk homeostasis theory in a simulated alarm handling task. Risk homeostasis theory predicts that, at a population level, the target level of risk remains constant. This prediction is rather disconcerting for the human factors community. Put plainly, it posits that following a human factors intervention, such as the reduction of alarm presentation rate to 15 messages per minute, the operators would engage in behavioural adjustments to return the environment to the previous level of risk. If people do indeed engage in this kind of compensation following risk reduction strategies it would appear to negate the effectiveness of any improvements made. In the study presented within this chapter Tom and Neville examine the extent to which homeostasis occurs in a situation which simulated physical risk. The findings of the study provide evidence against risk homeostasis theory. It would seem likely therefore, that human factors interventions made to alarm systems would make tangible improvements in the reduction of risk.

2

Urgency mapping in auditory warning signals

Judy Edworthy

Introduction

There are many occasions where neither a visual nor a verbal warning can be relied upon to attract a person's attention. For example, if a pilot is carrying out a tricky manœuvre he or she may not notice that one of the many visual displays is showing a dangerously low height, or that the fuel level is lower than anticipated. A nurse, attending a patient in an intensive care ward, cannot have his or her attention drawn towards a second patient in the ward by a spoken warning, first because this would be unethical and second because a voice warning may not be detected amongst all the other verbal messages being conveyed in the ward.

Thus there may be occasions where only non-verbal auditory warnings can be relied upon to get the operator's attention, and for this reason auditory warnings are widely used in aviation, hospitals and throughout industry (Thorning and Ablett, 1985; Kerr, 1985; Lazarus and Hoge, 1986). However, the traditional problems associated with such warnings are legion and, in many ways, these sort of warnings have become victims of their own success. Typically, such warnings, which have traditionally been bells, hooters, buzzers and the like, have been installed on as 'better–safe–than–sorry' principle (Patterson, 1985). Consequently they are too loud, too insistent, and tend to disrupt thought and communication at the very time that this is vital (e.g. Kerr and Hayes, 1983; Rood, Chillery *et al.*, 1985). For some situations, this may be a positive feature, especially if the desired response is that the person hearing the warning is to leave the vicinity in the fastest time possible, without the need to communicate. As an example, home fire alarms are typically shrill and piercing. If these acoustic attributes wake up the householder and cause him or her to leave the room in which the alarm has sounded at the earliest opportunity, then this must be a useful feature of that particular warning, no matter how aversive it may be. The same is true of burglar and car alarms.

It is necessary, therefore, to draw a contrast between the types of alarms which are intended to scare, such as car and burglar alarms, and those which are intended to inform, such as ones used in many high-workload environments. When a pilot or a nurse is carrying out a complex task, the last thing he or she may want is a shrill, piercing and insistent alarm; a signal which informs would be much more useful.

Another problem typically associated with such warnings is that there are usually too many of them in the environments in which they are used (Thorning and Ablett, 1985; Kerr, 1985; Montahan, Hetu *et al.*, 1993). A single patient in a multi-bedded intensive care ward may be monitored by several pieces of equipment, each of which can produce several different alarms. During our work we have observed more than 30 alarms to be associated with a single patient, and this takes no account of pieces of equipment that may be monitoring other patients in the ward.

These related problems of insistence, loudness and number of auditory warnings render many existing alarm systems effectively useless. The problems no doubt arise because of the fear of missed warnings. A signal that is not loud enough will go undetected, so an alarm might typically be made too loud for its environment (Patterson, 1982). Equally each situation, however unlikely it is to arise, needs to be signalled by some sound or other in case an accident, or litigation (or both) were to follow. Thus in terms of signal detection, most operators and manufacturers will probably feel safer with a high level of false positive responses than with potentially missed targets. Thus it is easy to see the source of the problem but harder to see a solution. However, 12 warnings are not necessarily twice as effective as six, and there is some chance that they are less so. The cognitive capacity and ability of the human operator, at the centre of nearly every system, are the most important features and therefore must be catered for in auditory warning design. Many psychological issues can come into play in proper, ergonomic, auditory warning design. This chapter focuses on some of the most pertinent of these in order to show how the appropriateness of auditory warnings signals might be improved. Several projects carried out over the last few years in the Department of Psychology, University of Plymouth will be reviewed. The focus will largely be on the psycho-acoustic and psychological, rather than the acoustic, aspects of warning design.

Issues in auditory warning design

Appropriate levels

Ensuring that a warning signal is neither too loud nor too quiet – in other words, that it is reliably detectable – has largely been solved in theory, although practice generally lags far behind. The solution lies in complex computer programs based on the theoretically predicted functioning of the auditory filter, which allows accurate prediction of masked threshold across a large range of frequencies. Examples of the use of such models and programs are the application of the masking model developed by Patterson (1982) in the design of auditory warnings for helicopters (Lower, Wheeler *et al.*, 1986) and the application of a slightly different model (Laroche, Tran Quoc *et al.*, 1991), based on a different model of the auditory filter, in the prediction of masking in intensive care wards and operating theatres (Momtahan, Hetu *et al.*, 1993). Some of these programs can take account of hearing loss with age (presbyacusis) and even individual hearing loss, if necessary.

The psychological and psycho-acoustic attributes of warnings, however, are less well documented and have been the main focus of our research at the University of Plymouth.

Psychological issues

In order to ascertain what might be important, psychologically, about auditory warnings, let us take a slightly different perspective on the problems already described. Why are warnings generally too insistent and aversive? Why does a manufacturer place an irritating alarm on a piece of equipment instead of a sensible, attention-getting but tolerable sound? An alarm that is shrill and irritating is conveying to the listener that the situation which it is signalling is urgent and that attention must be paid immediately to that situation. The manufacturer is telling the hearer that his or her warning must be heeded, as any responsible manufacturer would be expected to do. Unfortunately, if one is working in an environment in which all such situations are signalled by equally urgent warnings the effect may be lost. It may be that some situations requiring an alarm (or some kind of alerting sound) are not very urgent; therefore an implicitly urgent warning sound is not always appropriate. For example, a study by O'Carroll (1986) showed that, of 1455 soundings of alarms in a general purpose intensive therapy unit over a three-week period, only eight of these signalled potentially life-threatening problems. Consider too nurses working in neonatal units. They might wish just to be told of a problem – they may not want alarms to shout it at them or indeed the babies in the unit. This is true of many hospital environments, and this now seems to be filtering through to manufacturers. On the other hand, a newly-designed set of auditory warnings might be accepted or rejected by the user group simply on the basis that they are 'not urgent enough', or that the warnings for some situations are inappropriately urgent.

A study highlighting the urgency problem has been carried out by Momtahan and Tansley (1989) at a hospital in Ottawa, Canada. In this study, medical staff were asked to estimate how many alarms they would recognize of the 20 or so alarms typically heard in an operating and recovery room in which they often worked. In many cases, staff overestimated the number they believed they could recognize, and in some cases they could recognize only four or five of the warnings. In the second part of the study, these same subjects were asked to rate the medical urgency of each of the situations for which the warnings were designed. That is, the medical urgency of the situations were assessed independently of the warning sounds themselves. A second group of subjects were then asked to rate the acoustic urgency of each of these warnings. These subjects did not know the purpose of the alarms and so based their judgements entirely on the acoustic properties of the alarms. The results show, not surprisingly, that the two judgements were not correlated. In practice, this means that urgent medical situations are being signalled by alarms that may not be psycho-acoustically urgent, and vice versa. Clearly, a better match between situational and acoustic urgency would improve the work environment. Momtahan and Tansley refer to this matching as 'urgency mapping'.

Urgency mapping at first sight may seem to be rather a luxury; it could be argued that if the meaning of a warning is known, then the psycho-acoustic urgency of the alarm is unimportant. However, a more recent study (Momtahan, Hetu *et al.*, 1993) confirms that medical staff working in specific medical environments fail to identify accurately many of the warning sounds available in that environment. Urgency mapping, then, can help in the judgement of the urgency with which one should react, even if the precise meaning of the warning is not known.

In our studies we have largely been concerned with urgency as an attribute of sound, with a view to improving urgency mapping for situations where the meaning of the warning sound may not be known. One needs only to consider for a moment the area of animal calls to convince oneself that urgency is an attribute of many sounds and is, therefore, potentially amenable to investigation. A field in which sound is used effectively to manipulate urgency, amongst many other psychological variables, is that of film music. Here some of the effects are by association, but it is clear that many responses to such music are, for want of a better word, 'intuitive' (as an example, try to recollect the first time that you watched the Hitchcock film 'Psycho', and were perhaps unaware of the plot. Were you not on the edge of your seat even during the title sequence, which is accompanied by one of the most effective scores ever written?).

The psychological correlate of urgency, therefore, figures large in the design and implementation of non-verbal auditory warnings; the desire to convey urgency, whether appropriate or not, is probably one of the reasons why traditional warnings tend to be irritating and aversive; it is also a strong enough element in a set of newly-designed warnings to warrant the success or failure of that set. It would be useful if, therefore, it could be properly implemented into auditory warning design. In order to do this, quantifiable data are required. The next section of this chapter describes this empirical evidence in detail.

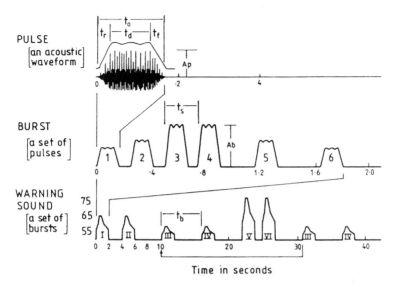

Figure 2.1   Patterson's prototype for warning construction.

Experimental studies

Design guidelines

If one is restricted to the use of bells, horns, buzzers and so on it is difficult to manipulate the urgency of a warning over any great psychological range; however, a set of guidelines for auditory warning production devised by Patterson (1982) allows much greater manipulation of this dimension. Patterson's guidelines show how appropriate loudness levels can be predicted for warnings intended for complex noise environments (the guidelines were specifically produced for civil aircraft, but could be applied in many work environments because they focus on psycho-acoustic, rather than environment-specific, variables) and how warnings can be designed to be ergonomically tailored to that environment. The method of construction of these warnings needs to be described in detail, because it forms the basis of many of our studies. However, our findings can be applied more or less across the range of non-verbal auditory warnings to a greater or lesser extent. Patterson's guidelines present us with a useful methodological framework in exploring the nature of perceived urgency.

Once the appropriate loudness level for warning components has been established, the warning itself is constructed in three stages (Figure 2.1): first, a small unit of sound, a pulse, is designed. This unit carries all the acoustic information needed for pitch determination, localization and so on. The second element is the burst of sound, which is produced by playing the pulse several times. The pulses that make up this burst need not have the same pitch, or the same loudness level, or even have regular time intervals between each of them. The basic harmonic information in the pulse, however, is kept constant so that the burst has the same timbre throughout. In other words, the burst is somewhat akin to a short melody played on a single musical instrument. The burst would typically last about two seconds. The final stage of warning construction is that of the complete warning, where the burst is played once or twice, followed by silence, followed by other bursts. This would continue until the situation which the warning is signalling is alleviated. The burst may come on at different levels of urgency; if the situation has a high priority, a more urgent version may follow the initial sounding of the burst. This more urgent version might be louder, faster, and at a higher pitch than the initial burst. If the situation is not urgent, then a 'background' form of the burst might be heard, which could be quieter, slower, and possibly at a lower pitch than the initial burst.

Patterson's guidelines thus demonstrate not only a more ergonomic way of constructing auditory warnings, they also begin to suggest ways in which the urgency of warning could be manipulated. Our studies have taken the exploration of urgency much further than this, and in some cases we have attempted to quantify formally the relationship between the subjective, psychological construct of urgency, and the objectively measurable acoustic parameters which convey that urgency.

Experimental studies I: ranking studies

The method of construction advocated by Patterson introduces a number of acoustic parameters which may affect the urgency of the resulting auditory warning. Some of these parameters affect the harmonic quality of the warning (the pulse) and some affect the melodic and temporal quality of the warning (the burst). In the first major study systematic and detailed observations of the effects of both types of parameters were carried out, and on the basis of these results predictions were made about the urgency of warnings generated from combinations of these parameters (Edworthy, Loxley *et al.*, 1991).

Pulse parameters

A pulse, in common with almost all other sounds, is made up of several harmonic components which give the sound its distinctive timbre. For instance, the particular combination of harmonics is the feature which distinguishes the same note, at the same loudness, when played by a clarinet or by a flute. It is likely that certain features of this harmonic content also affect perceived urgency. As well as having harmonic features, pulses of sound also possess temporal features, one of the most important of which is the way in which the amplitude of the sound progresses throughout its duration. That is, whether the sound reaches maximum output quickly, like a plucked sound, or slowly,

Table 2.1   Stimuli, pulse experiments

Parameter	Levels
Fundamental frequency	150, 200, 350, 530 Hz
Harmonic series	Regular, 10% irregular, 50% irregular, random
Delayed harmonics	Present, absent
Amplitude envelope	Slow onset, regular, slow offset

Table 2.2   Direction of effects, pulse experiments

Parameter	Direction of effect
Fundamental frequency	High > low
Harmonic series	Random/10% irregular > 50% irregular > regular
Delayed harmonics	No delayed harmonics > delayed harmonics
Amplitude envelope	Regular/slow onset > slow offset

Key: > More urgent than
　　/ Equally urgent

like a bowed sound. This feature, referred to as 'amplitude envelope', was also explored in these experiments. Table 2.1 describes the parameters tested in the first series of experiments.

The fundamental frequency of the pulse usually dictates the pitch of the pulse (but not always); thus, the higher the frequency, the higher the pitch. The regularity of the harmonic series affects the ability of the listener to attach a particular pitch to a stimulus; generally, the more irregular the harmonic series, the more difficult it becomes to do this. The other general perceptual effect of increasing inharmonicity is to make an acoustic stimulus more rasping, and less musical-sounding and pure. Delayed harmonics occur when some of the harmonics in the sound do not play immediately the pulse is heard but come on at some later point; this has the effect usually of introducing a small pitch change at the point where the extra harmonics come on. The final manipulation carried out on the stimuli was the imposition of one of three types of amplitude envelope to the pulses; a slow onset, where the pulse reaches maximum intensity level slowly (relative to the overall length of the pulse), a slow offset, where the pulse reaches maximum intensity quickly, but tails off slowly for the rest of the pulse (the inverse of slow onset), and a 'regular' envelope (defined here for operational purposes only) which has the quicker of the two available onset and offset envelopes. This stimulus reaches maximum intensity quickly, and sustains this level for the whole of the pulse, until the final quick offset. Fuller details, plus detailed diagrams of these manipulations, can be seen elsewhere (Edworthy, Loxley *et al.*, 1991).

The results showed that most of our intuitive notions about how perceived urgency might change as a result of these manipulations were borne out, although there were some surprising effects. The results are shown in Table 2.2.

Table 2.3   Stimuli, burst experiments

Parameter	Levels
Speed	Fast, moderate, slow
Number of repeating units	4, 2, 1
Rhythm	Regular, syncopated
Speed change	Slowing, regular, speeding
Pitch contour	Down/up, random
Pitch range	Small, moderate, large
Musical structure	Resolved, unresolved, atonal

Thus increases in both fundamental frequency and harmonic irregularity increase the urgency of pulses. The effect for inharmonicity was more or less as could have been predicted, with urgency increasing generally as the ratio between the fundamental frequency and the harmonic series above it increases in complexity. A random harmonic series produces the most complex ratio, followed by 10 per cent inharmonicity, followed 50 per cent inharmonicity, followed by a regular harmonic series, for which the ratio is 1:1. The presence of delayed harmonics appears to decrease urgency, and this was one of the first indications that urgency is a separable psychological dimension from distinctiveness; the presence of delayed harmonics in a pulse certainly makes it distinctive, but the results show that they decrease its urgency relative to the same pulse with no delayed harmonics. Along the temporal dimension it was found that a slow onset envelope was judged to be more urgent than a fast onset, which was somewhat surprising. This was interpreted in terms of a slow onset resembling an approaching object, which might in turn generate a sense of urgency.

One of the most interesting general findings from this first set of experiments was that participants readily responded to the task itself and were remarkably consistent in their responses. The method of investigation used allowed us to measure the degree of consistency shown by subjects, which was very high in general. It was also apparent from these studies that some parameters produce greater effects on perceived urgency than others. This was not unexpected, but the method of investigation used in these studies did not allow us to measure the relative strengths of these effects. This line of investigation is, however, followed in other experiments (Hellier and Edworthy, 1989; Hellier, Edworthy *et al.*, 1993a; Hellier, Edworthy *et al.*, 1993b). These experiments will be described later in this chapter.

Burst parameters

In the second set of experiments associated with this project, the effects of a range of melodic and temporal parameters on urgency were explored. These are the parameters which affect the urgency of the burst of sound. Table 2.3 lists in detail the parameters explored in this set of experiments.

Table 2.4 Direction of effects, burst experiments

Parameter	Direction of effect
Speed	Fast > moderate > slow
Number of repeating units	4 > 2 > 1
Rhythm	Regular > syncopated
Speed change	Speeding up > regular/slowing
Pitch contour	Random > down/up
Pitch range	Large > small > moderate
Musical structure	Atonal > unresolved > resolved

Key: > More urgent than

The burst parameters investigated covered several aspects of both the temporal and the melodic patterning of the burst. The speed of the burst was determined by the inter-pulse interval, with faster bursts possessing shorter inter-pulse intervals. Speed change was introduced into the stimuli by either increasing, decreasing or maintaining a fixed inter-pulse interval. A regular rhythm was obtained by maintaining a constant inter-pulse interval, and a syncopated rhythm was achieved by alternating a short with a long inter-pulse interval between each successive pulse. The 'number of repeating units' dimension was achieved by playing a short, four-pulse unit of sound once, twice or four times. Inevitably, this dimension also affects the length of the resultant burst.

The other group of parameters investigated in this set of studies were melodic in nature. The pitch contour refers to the sequence of ups and downs in the pulse, and for the purposes of these experiments just two of many possible contours were investigated; a down/up contour, where the burst possessed one change of direction during its progression, and 'random', where as many contour changes as possible were introduced into the burst. Pitch range refers to the frequency ratio between the highest and lowest frequencies to be found in the pulse, so that with a small pitch range this ratio was small (in fact, four semitones) and for the large pitch range, it was much greater (10 semitones in this study). The final melodic dimension investigated was musical structure, where the three levels reflected the perceived musical finality and completeness of the burst. For a resolved burst, the resulting sound is perceptually complete; for an unresolved burst, this is not so, but the burst remains melodic; and for an atonal burst, the melodic structure is so obscured as to make the perception of resolution impossible because the designation of a tonal centre, the focus of the resolution, is impossible.

As in the earlier experiments, participants were very consistent in their urgency judgements. Almost every feature selected produced a significant effect on perceived urgency. Table 2.4 summarizes these effects.

Most of the temporal parameters produced the expected effects on perceived urgency; faster, speeded up and more repetitious bursts produced greater estimations of urgency. Somewhat surprisingly, a regular rhythm produced higher estimations of urgency than a syncopated rhythm. However, as there are potentially many versions of rhythmic irregularity, it would be wise at this point to reserve judgement on this particular effect. Amongst the melodic parameters, pitch contour was found to have some effect (although the same reservation stated above must be reiterated here, as there are many pitch contours available for investigation), as did the musical structure, both in the directions that would be expected. Pitch range produced a more surprising effect, showing a non-monotonic function. One interpretation of this effect is that small pitch ranges introduce an element of chromaticity into the structure (with reference to a piano, this would mean using a small cluster of both black and white notes) which increases its urgency relative to one where a cluster of diatonic notes (the members of the usual scale) might be used, for example, in a burst with a more moderate pitch range.

As before, we also found that some parameters appeared to produce greater effects on perceived urgency than others, and some further investigation was again necessary using different methods. In general, we established that temporal features produced greater and more consistent effects than melodic features. This is not to say, though, that temporal parameters are more important than melodic parameters because the levels of the parameters chosen for investigation may have covered a greater perceptual range than the melodic parameters.

As a result of these two sets of experiments, we now have a database showing the direction of the effects on perceived urgency of most of the important parameters at our disposal in auditory warning design. Two important parameters are missing, and for good reason; first, we excluded loudness because, for any given environment, the loudness of a warning must be fixed within a fairly narrow range in order to meet the dual requirements of being audible and not too loud. Loudness, however, does have very clear effects, as similar studies on perceived urgency by Momtahan (1990) show. Another feature not investigated was the distribution of the harmonics; it would be expected that a pulse weighted more with higher harmonics would be perceived as being more urgent than one in which all harmonics are equally weighted, or where the harmonics are generally lower. Momtahan's study again shows this to be the case, although there is another study by Halpern, Blake *et al.* (1986) which makes rather a different claim about the irritating qualities of certain sounds. We chose not to investigate this parameter because, again, it is of little use in auditory warning design where the pulse has to be harmonically tailored to the noise environment. The noise spectrum will dictate how individual harmonics are to be weighted, so it will not normally be possible to weight the harmonics in other ways. Rather, there are better ways to manipulate perceived urgency. Aside from these two missing parameters, our database covers most of those parameters likely to be of practical use.

Predicting urgency in warnings

In the final part of this large study, pulse and burst parameters were combined in order to construct acoustic stimuli which sounded much more like typical warnings, and specific predictions were made about their relative urgency. We constructed pulses predicted to have high levels of urgency on the basis of the pulse experiments, and then used these pulses to create bursts also predicted to have high levels of urgency. Bursts predicted to have low levels of urgency were also constructed. Altogether, 13 warnings were constructed and the order of urgency of this set were predicted prior to testing (for details of the warning specifications, Edworthy, Loxley *et al.*, 1991). Two important results emerged from this study; the first was that participants were remarkably consistent in their judgements, even more so than they had been for the earlier experiments. That is they had no problems either in assessing the relative urgencies of each of the warnings, or in maintaining a particular rank ordering for the entire experiment. This provides further evidence that urgency is a salient and meaningful psychological construct for auditory warnings. The second result was that the correlation between the predicted and the obtained rank ordering was highly significant. Apart from one or two slight displacements, the rank ordering of the 13 warnings obtained by experimentation was the same as the rank ordering that had been predicted. This result therefore validated our earlier results and our assumptions about perceived urgency.

Experimental studies II: psychophysical studies

The experiments reported above show only the effects of individual sound parameters on urgency; they do not show the relative strengths of individual parameters, not do they show whether some parameters behave in a more systematic and predictable way than others. It is useful to know this not only from a research point of view, but also from a practical design viewpoint. For example, if a manufacturer has developed a warning and has been told that it must be modified because it is too urgent, or not urgent enough, what might be the most efficient way of altering its urgency? Just a small change in the warning speed may do the trick, whereas a large change in an alternative parameter might be necessary to effect the same change. In order to find out about the relative strengths of individual sound parameters on perceived urgency, a different methodological approach was taken to the study of urgency by exploring the effects using psychophysical methods.

The basis of psychophysics is that there is a measurable relationship between the amount of physical sensation and the subjective experience of that sensation. Furthermore, the relationship between the two can be quantified and scaled. Most traditional psychophysics have sought to quantify the relationship between the objective level of various parameters found in the physical world, such as light, temperature, weight and so on, and the subjective sensation of that stimulus, for example, 'heaviness'. Increases in the level of the physical stimulus inevitably produce subjective increases in sensation. Stevens (1957) encapsulated this relationship between objective and subjective variables with the power law equation, which takes the general form

$$S = kO^m$$

where S is the subjective value ascribed to a stimulus, k is a constant, O is the objective, measurable value of the stimulus and m is the exponent, based on the slope of the graph generated when the subjective and objective values are plotted against one another. The larger the exponent, the greater the change in the subjective value per unit change in objective value. Some sensations have very high exponents, whilst others have exponents that are much lower (Stevens and Galanter, 1957). As many of the acoustic parameters which convey urgency can be objectively quantified, we proposed that the power law could also be used to help in the quantification of perceived urgency, with the exponents derived for individual sound parameters showing the strength of these individual effects.

Some of the parameters tested in our earlier work are readily quantifiable, and so lend themselves to investigation using psychophysical techniques. Parameters such as speed and fundamental frequency can be quantified. Others, such as rhythm, musical structure and so on cannot, and are clearly excluded from such investigations.

Deriving urgency exponents

In the earliest psychophysical studies, it was established that the power law was suitable for application in this area (Hellier (1991); Hellier, Edworthy *et al.* (1993a)). Our experiments confirmed that free magnitude estimation, along with line length, were the most accurate methods of measuring subjective perceived urgency and hence the methods most likely to yield accurate exponent values.

A series of studies was then carried out using these techniques on some of the more important and quantifiable parameters known to affect perceived urgency (Hellier, Edworthy *et al.*, 1993b). Over a series of four experiments, speed, pitch, number of units repetition and degree of inharmonicity were investigated, and the relationship between objective, measurable changes in these four parameters and subjective ratings of urgency was charted. The values of the exponents obtained for each of the four parameters can be seen in Table 2.5.

The larger the exponent, the greater is the change in subjective level (urgency) produced by a unit change, or percentage change, in the physical parameter. Thus greater changes are produced by, for example, a 50 per cent increase in speed than by an equivalent increase in any of the other three

Table 2.5 Urgency power functions

Parameter	Exponent
Speed	1.35
Fundamental frequency	0.38
Number of repeating units	0.50
Inharmonicity	0.12

parameters. In fact, a 50 per cent increase in urgency could be brought about by a 1.3-fold increase in speed, a 2.8-fold increase in fundamental frequency, a 2.2-fold increase in the number of units of repetition, and by a massive 28.5-fold increase in inharmonicity. The very low value of the inharmonicity exponent renders this parameter practically useless for the manipulation of urgency. The other three, however, are potentially very useful and some of these applications will be discussed later in the chapter.

In all of these studies, we found that the data provided a good fit to a straight line; in other words, quantification of this sort proved meaningful, as had also been indicated by our earlier studies where we had simply looked at the effects of individual parameters on perceived urgency. Quantification has useful applications, because it allows for prediction and recommendation. However, it also allows other, rather more theoretical assumptions to be tested, and it was this route that we explored next. Using the data obtained in the first four experiments, it was possible to predict equally urgent sounds and equal changes in urgency across different parameters, something that was not possible from the ranking studies described earlier.

Equalizing urgency across acoustic parameters

In the final experiment in this study we selected three theoretically equal levels of urgency for each of three parameters – fundamental frequency, speed and number of repeating units which were labelled 'high', 'medium' and 'low'. These three levels were obtained by selecting three values, or levels, of urgency, and by using the exponent to calculate the levels of the acoustic parameters required to produce these levels of urgency. We combined these three levels in such a way as to generate every possible stimuli – 27 in total. Thus one stimulus consisted of all three high levels, three stimuli consisted of two high and one medium level and so on. Subjects ranked the urgency of each of these stimuli, and the expected rank order was predicted beforehand, taking account of the premise that equal units and equal changes in urgency should be equivalent despite being conveyed through different parameters. This predicted correlation proved significant when compared with the obtained order. Thus our prediction, namely that equal levels and equal changes in urgency can be brought about by different parameters if the exponents are obtained and then interpreted carefully, was borne out.

However, it was also found that one of the parameters, fundamental frequency, had a greater influence on urgency than the others. This was confirmed when multiple regression was carried out on the data. So, although the stimuli employed theoretically equal units of urgency between the three parameters, there were indications that pitch had a greater influence than the other parameters. A number of reasons for this are possible and are discussed in detail elsewhere (Hellier, Edworthy *et al.*, 1993b).

Design implications

First and foremost, our studies have shown that perceived urgency in sound, as a subjective attribute, can be readily and reliably measured using two different experimental techniques, one based on rank ordering and the other on magnitude estimation and subsequent quantification. This suggests that people find it easy to respond to questions about perceived urgency which implies in turn that it is a salient feature of the kinds of sounds tested in our experiments. Thus it is likely also to be a salient feature of auditory warnings, previously suggested only by anecdotal evidence. Our studies also show that it is possible to predict the urgency of warnings on the basis of knowledge about their acoustic structure. Furthermore, the quantification studies indicate that the relationship between at least some acoustic parameters and perceived urgency can be described accurately in terms of Steven's power law. These results tend to confirm and add precision to the earlier findings, but they also allow prediction. The greater the value of the exponent, the more economical the parameter is in producing changes in urgency. This information could be used in producing warnings sets where differences in urgency are required, or in advising manufacturers on the most efficient ways of adjusting the urgency of the warnings that they already use. The results can also be used to produce warnings which are different from one another, but which might have approximately equal levels of urgency, another important practical requirement.

Other psychological features of warnings

Urgency, clearly important in auditory warning design work, is not the only attribute of a warning; bursts of sound can convey a whole range of other meanings as well. For example, a particular sound might convey that an object is falling, that it is light, heavy, confused, rushed or a whole range of other meanings. The relative importance of these meanings has been explored in another of our group's projects (Loxley, 1992). We have investigated the relationship between sound parameters and a large set of descriptors in order to design sets of trend monitoring sounds which function somewhat like warnings, but which are intended for use as auditory feedback during, for example, difficult flying manœuvres. In sounds of these sort, other meanings as well as urgency are important. For instance, these sounds can mimic the trend being monitored through the sound parameter(s) chosen to convey that trend, provided all the possible meanings of that parameter are known. The problem becomes particularly interesting psychologically, though tiresome practically, when the use of a particular feature of a sound parameter conveys two or more conflicting meanings. One of the best examples of this is the use of a falling pitch pattern; this can convey both decreasing urgency and the sense of an object falling; if this pattern is used to convey a slowing helicopter rotor speed, the pilot may be being told simultaneously that his or her helicopter is falling out of the sky, and that the situation is becoming less urgent! Having completed our study of the meanings, in addition to urgency, of the most important sound parameters in use in warning and trend monitoring sound ('trendson') design, we are currently addressing the problem of contradictory information in sound. This work will add substantially to our growing database on the meaning of sound and its application.

References

Edworthy, J., Loxley, S. and Dennis, I., 1991, Improving auditory warning design: relationship between warning sound parameters and perceived urgency, *Human Factors*, **33** (2), 205–31.

Halpern, D., Black, R. and Hillenbrand, J., 1986, Psychoacoustics of a chilling sound, *Perception & Psychophysics*, **39** (2), 77–80.

Hellier, E.J., 1991, 'An investigation into the perceived urgency of auditory warnings', unpublished PhD thesis, Polytechnic South West, Plymouth.

Hellier, E. and Edworthy, J., 1989, Quantifying the perceived urgency of auditory warnings, *Canadian Acoustics*, **17** (4), 3–11.

Hellier, E., Edworthy, J. and Dennis, I., 1993a, A comparison of different techniques for scaling perceived urgency, *Ergonomics*, (in press).

Hellier, E., Edworthy, J. and Dennis, I., 1993b, Improving auditory warning design: quantifying and predicting the effects of different warning parameters on perceived urgency, *Human Factors*, (in press).

Kerr, J.L., 1985, Auditory warnings in intensive care units and operating theatres, *Ergonomics International 85*, 172–4.

Kerr, J.H. and Hayes, B., 1983, An 'alarming' situation in the intensive care ward, *Intensive Care Medicine*, **9**, 103–4.

Laroche, C., Tran Quoc, H., Hetu, R. and McDuff, S., 1991, 'Detectsound': a computerised model for predicting the detectability of warning signals in noisy workplaces, *Applied Acoustics*, **32** (3), 193–214.

Lazarus, H. and Hoge, H., 1986, Industrial safety: acoustic signals for danger situations in factories, *Applied Ergonomics*, **17**, 41–6.

Lower, M., Wheeler, P., Patterson, R., Edworthy, J., Shailer, M., Milroy, R., Rood, G. and Chillery, J., 1986, The design and production of auditory warnings for helicopters 1: the Sea King, *ISVR Report No AC527A*.

Loxley, S.L., 1992, 'An investigation of subjective interpretations of auditory stimuli for the design of trend monitoring sounds', unpublished MPhil thesis, Polytechnic South West, Plymouth.

Momtahan, K.L., 1990, 'Mapping of psychoacoustic parameters to the perceived urgency of auditory warning signals', unpublished Master's thesis, Carleton University, Ottawa, Ontario, Canada.

Momtahan, K.L. and Tansley, B.W., 1989, 'An ergonomic analysis of the auditory alarm signals in the operating room and recovery room', presentation at the Annual Conference of the Canadian Acoustical Association, Halifax, Nova Scotia.

Momtahan, K.L., Hetu, R. and Tansley, B.W., 1993, Audibility and identification of auditory alarms in operating rooms and an intensive care unit, *Ergonomics*, (in press).

O'Carroll, T.M., 1986, Survey of alarms in an intensive therapy unit, *Anaesthesia*, **41**, 742–4.

Patterson, R.D., 1982, Guidelines for auditory warning systems on civil aircraft, CAA paper 82017, (London, Civil Aviation Authority).

Patterson, R.D., 1985, Auditory warning systems for high-workload environments, *Ergonomics International 85*, 163–5.

Rood, G.M., Chillery, J.A. and Collister, J.B., 1985, Requirements and application of auditory warnings to military helicopters, *Ergonomics International 85*, 169–72.

Stevens, S.S., 1957, On the psychophysical law, *Psychological Review*, **64**, 153–81.

Stevens, S.S. and Galanter, E., 1957, Ratio scales and category scales for a dozen perceptual continua, *Journal of Experimental Psychology*, **54**, 377–411.

Thorning, A.G. and Ablett, R.M., 1985, Auditory warning systems on commercial Transport aircraft, *Ergonomics International 85*, 166–8.

3

An experiment to support the design of VDU-based alarm lists for power plant operators

P.D. Hollywell and E.C. Marshall

Introduction

Background

Control room operators use large arrays of visual information when monitoring and controlling power station plant. Increased use is being made of computer-driven visual display unit (VDU)-based displays in control rooms and care must therefore be taken to ensure that they are designed to present a large amount of information effectively. The enormous amount of data now being presented to operators is exemplified in the nuclear power industry.

A modern UK twin advanced gas-cooled reactor (AGR) has typically about 2500 analogue and 3500 digital inputs per reactor going to the control room data processing system (DPS). Plans to update the computer systems of the earlier UK AGR stations will result in systems typically with 4000 analogue and 5000 digital inputs per reactor. This information will be presented to two operators and a supervisor in the control room via VDU formats numbering between 200 and 300, with additional hard-wired display panels (Jackson, 1988). For the UK Sizewell 'B' pressurized water reactor (PWR) it is anticipated that there will be between 10 000 and 20 000 separate inputs from plant sensors and perhaps 30 VDUs, with additional hard-wired consoles monitored by three or four operators in one room (Singleton, 1985).

Alarms are distinct signals which are displayed to attract the control room operator's attention to abnormal plant conditions. Traditionally, alarms have been presented as messages on illuminated tiles (annunciators) mounted in the control panels, accompanied by an audible signal such as a bell or tone.

The advent of computer technology has enabled the VDU presentation of alarms either as text messages or as highlighted symbols on mimic diagrams. This has allowed plant designers to include many more process alarms with more detailed messages than was possible with annunciator panels.

The rapid onset of large numbers of alarm text messages during a plant disturbance can make it difficult for the operator to read and understand all the necessary information for performing a rapid and accurate diagnosis of process state. Hence, in order to design an adequate computer-based system, it is important to know the rate at which operators can assimilate alarm information from a VDU screen. Although much generic research into alarm systems has been carried out, no reported experiments have ever directly addressed the fundamental question of the maximum rate at which operators can read text messages from a VDU-based alarm list.

Alarms

An alarm is the consequence of a parameter (typically temperature, pressure or voltage) exceeding the limits, specified by the designer or the operators, for the normal operation of the plant. An alarm is often accompanied by visual and/or audible cues (e.g. lights and bells) to ensure that it claims the attention of the operators (Singleton, 1989). In several UK nuclear power station control rooms the majority of alarms are presented as text messages listed on a VDU screen (i.e. VDU alarm list), though some annunciators still remain. At Heysham 2, a recently commissioned UK AGR, there are about 18 000 alarms per reactor (Jackson, 1988). A VDU alarm list typically consists of a list of one line messages which appear on the screen in time sequence. A new alarm appears at the bottom of the list and, as the screen fills after 25 or so messages, the screen is cleared and an alarm message appears at the top of a new page. Various facilities may be provided for paging back and forth through the list and, in addition, highlighting techniques such as colour coding may be provided to distinguish between different information types.

World-wide experience shows that during a plant incident operators can sometimes be presented with a large number of alarm messages in a very short period of time. During one loss-of-coolant incident at a US nuclear reactor, more than 500 annunciators changed status within the first minute and more than 800 within the first two (Sheridan, 1981). Increased concern over the design and operation of computer-based alarm systems was raised by the Three Mile Island incident in the US in 1979 where nuclear power station control room operators were overwhelmed by the sheer number of alarms and this was cited as the main reason why operators overlooked crucial alarms.

> During the first few minutes of the accident, more than 100 alarms went off, and there was no system for suppressing the unimportant signals so that operators could concentrate on the significant alarms...

> Kemeny (1979)

Most of the blame for the accident was attributed to poor control room layout, confusing alarm systems and poorly trained operators (Milne, 1984). Problems at UK plants and elsewhere have shown that difficulties with computer-based alarm systems are not restricted to US nuclear stations. In June 1984, operators at Heysham 1 nuclear power station failed to respond when data and alarm displays registered higher than average temperatures in reactor-fuel channels. Alarms showing that two channels were registering higher than average gas outlet temperatures were missed by the operators. It was half-an-hour later, when a computer-assisted scan highlighted the abnormal temperatures, that checks were finally made and the reactor was shutdown manually. The original alarms appeared on two VDU alarm lists on the reactor operator's desk. A report by the Nuclear Installations Inspectorate concluded that the event 'revealed weaknesses in procedure, ergonomics and human performance' in the control room (Milne, 1984). Marshall (1990) confirms that alarm arrival rates can be very high during transients at AGR stations, and alarm rates in excess of 100 per minute could be sustained for up to three minutes.

It should be noted that high numbers of alarms exist in all types of power stations, process industries and in the aerospace industry (Singleton, 1989). Since it is likely that during incidents operators could be presented with a large number of alarms in a short space of time, problems with computer-based alarm systems could also be prevalent in these industries. This is especially true with the increasing acceptance of the VDU alarm list as a *de facto* standard in most of these industries.

Alarm systems research

There have been few experiments which have directly assessed the rate at which operators can read text-based alarm messages. Recent experiments investigating aspects of computer-based alarm systems have tended to compare overall performance by operators when using different modes of alarm presentation; i.e. annunciators versus text lists (EPRI, 1988), annunciators versus colour mimics combined with alarm logic (Reiersen, Marshall *et al.*, 1987), evaluation of a proposed information display and alarm system (Hollywell, 1990). Although reading speed was not directly addressed in the Hollywell study, different alarm presentation rates were applied as an experimental condition. No real differences in performance were observed when alarm rates were doubled, though the presentation rates were low and for short duration. The above studies suggest that there are improved ways of presenting alarms other than alarm lists, in terms of speed of alarm detection, time taken to make diagnoses after alarm detection and accuracy of diagnosis.

The study of reading skill has provided a considerable body of literature in general cognitive psychology (Barber, 1988). More recently, applied human factors researchers have been concerned with the way VDU screen reading performance for blocks of text compares with that for a printed page. In general, the findings have suggested that reading text from a VDU tends to be slower, less accurate and more fatiguing (Oborne and Holton, 1988; Dillon, McKnight *et al.*, 1988; Gould, Alfaro *et al.*, 1987). However, these studies do not directly relate to process control alarm messages. Generally, these studies conclude that reading speed may be between 20 per cent and 30 per cent slower from a VDU screen than from the printed page; i.e. in the order of 180 words per minute rather than over 200. Accuracy depends on the measures used in the experiment, but it would seem that routine spelling checks are not affected by the presentation media, however performance deficits may occur with more demanding cognitive tasks, such as comprehension. As to why this happens, recent work suggests that it is due to the poor definition, low resolution and general illegibility of VDU-generated text and the high definition graphic displays now available (i.e. a resolution of at least 1000×800 pixels) should enable reading performance to be equal to that achieved with the printed page (Gould, Alfaro *et al.*, 1987).

In considering a process control operator reading alarm text messages from a VDU, it is clear that the task demands are clearly different from those faced by a subject with no distractions in a VDU text reading experiment. Typically, if a number of alarms appears on the screen, the operator should read each one in turn and identify if any necessary actions are required. If the operator is expected to access the defined procedure and carry it out, then dealing with a group of alarms is going to be a fairly slow process. If, on the other hand, the operator is only required to scan down the group of alarms and then interpret them in terms of a particular plant state or fault, then reading might be relatively fast. The latter is the kind of diagnostic task investigated in the alarm presentation experiments referred to earlier.

An alarm list screen is much more structured than plain text, sometimes with clearly defined columns for time, code, message, severity etc. The vocabulary is a restricted one and abbreviations familiar to operators are employed. This would suggest that reading a list of alarms could well be much faster than reading plain text. However, the legibility of the text, VDU resolution and lighting conditions in the control room may be less than optimum and, as the information may be crucial, the operator may be more careful in reading it.

Danchak (1988) showed the importance of structure in textual alarm messages. He used a VDU alarm presentation system to investigate performance differences in the way the alarm text messages were structured in terms of the field positions within the message. Each subject was instructed to look at a VDU screen, which displayed a static single alarm message, on hearing an audible tone. The subject was asked to press a button when they understood the message. The experimenters stressed the need for quickness and accuracy. Pressing the button blanked the screen and the subject had to duplicate the alarm message on a standard form. When the subject had completed this task the next alarm would appear, triggered at random times. Response times and accuracy were recorded, though accuracy was eventually discarded since error rates were virtually zero for all subjects. Mean response time for a single message was about four seconds. If this result is extrapolated for a continuous stream of alarms presented in a static single alarm message format with no time stress, then an operator should be capable of reading and understanding about 15 alarms per minute.

In a study by Halstead-Nussloch and Granda (1984), the effect of varying message presentation rates and task complexity on subject performance was investigated. The variables examined were message rate, message presentation format, probability of target message and number of response categories (1, 2 or 4) for target messages. The subject viewed a VDU screen on which a stream of single-line messages either:

1. started at the bottom of the screen and then continually scrolled up the screen with each successive message; or
2. started at the top of the screen and successive messages were written below their predecessors, until the bottom of the screen was reached and then new messages overwrote the old messages starting at the top of the screen.

Each simple message consisted of a subject, a verb and an object. The messages were divided into target and non-target types. Accuracy was used as the measure of task performance.

The results from the experiments showed that increasing the message rate and the number of categories significantly reduced the accuracy of performance, while changes in presentation format did not. The authors claimed that at message rates of 60 messages per minute and less, subject performance was resource-limited; i.e. the subjects had enough time to read and understand the messages, but not necessarily enough time to accurately respond to them. At message rates above 60 messages per minute, subject performance was data-limited; i.e. the subjects did not have enough time to read and understand the messages, let alone have enough time to respond accurately to them. Results indicated that with a fixed target probability of 90 per cent, 30 alarm messages per minute produced an accuracy of about 98 per cent and 60 alarms per minute produced an accuracy of about 90 per cent.

Marshall (Hollywell, 1990), estimated that an operator was capable of reading and interpreting between 5–10 alarm messages per minute. However, he pointed out that the task context is crucial when attempting to quantify operator alarm reading performance. Singleton (1989) agreed that it is important to know what the operator was required to do with the alarms but, with that proviso, suggested that about five alarms per minute would seem a reasonable maximum alarm reading rate.

Objectives of experiment

The UK nuclear power industry has recognized the problems that may arise when an operator is required to deal with large numbers of alarms presented at a rapid rate. It has felt that it was important to design VDU-based alarm systems taking full account of the performance limits of control room operators. As part of on-going developments, an experiment was commissioned to assess systematically the rate at which nuclear power plant operators could read and identify typical alarm text messages presented on VDU alarm lists. The objective of the experiment was therefore to provide information on:

- the maximum rate at which operators can read alarm messages;
- individual operators' subjectively preferred rates of reading alarm messages;
- the nature of the operators' performance degradation observed at increasing alarm message presentation rates.

The results obtained from the experiment were thus intended to be of value to both VDU-based alarm system designers concerned with specifying total system performance and to ergonomists concerned with operator performance during incidents.

Experimental method

Experimental tasks

It was of prime importance to devise tasks which provided valid data in support of the aims of the study. That is, they needed to have an adequate degree of experimental validity. If experienced operators were to be involved in the experiments it was particularly important for the tasks to have an acceptable degree of face validity; i.e. they needed to appear to be testing alarm handling performance. For a detailed discussion of appropriate techniques for devising and selecting experimental tasks valid for the evaluation of process control interfaces, see Baker and Marshall (1986).

Four computer-based tasks were devised for this experimental study: a general knowledge test, a verbal reasoning test, an alarm categorization test and an alarm scenario categorization test. These four tests are briefly outlined below.

General knowledge test

Thirty simple general knowledge, true or false questions were asked. This task was used only to familiarize subjects with the computer test system and there was no analysis of the results.

Verbal reasoning test

This test was originally developed by Baddeley (1968). It provided a context-free test, similar to a simple alarm handling activity, for evaluating and assessing the computer test system and for future possible comparison of experienced operator and non-operator subject populations. The test consisted of presenting a written series of reasoning statements to a subject and then asking the subject to say whether the statement is true or false. The subject was required to do this for a series of 32 statements, as quickly and as accurately as possible.

This test was chosen because it is similar to the central alarm handling task and it is very simple to administer. It cannot be easily learned, so reducing practice effects. Also, it is claimed not to be fatiguing, thus minimizing any fatigue effects. If subjects only guessed in this test, the average number of successful responses would be 50 per cent.

Alarm categorization test

Alarm text messages from the three AGR power stations nominated for the study were used in this test. Subjects were told that the reactor had just tripped and that alarm system maintenance was in progress in a specified plant area. Subjects were required to assign each alarm message to one of three categories: expected, unexpected or maintenance. These categories were later noted to be similar to those used by Fujita (1989). This form of the test was used in order to ensure that the subject read the whole message string before categorizing the alarm.

In order that the alarm lists presented an acceptable degree of face validity for use with experienced operators, station-specific, scenario-independent alarms were used. However, alarm messages were presented in a random order in an attempt to prevent operators' previous training and operational experience, together with their exposure to particular scenarios, having an effect on their test performance.

Alarm scenario categorization test

This test was similar to the previous test in that subjects were required to categorize alarm messages as before. In this case, however, a sequence of alarm messages based on a station-specific scenario was used in an attempt to assess to what extent subjects' previous training and operational experience affected their test performance. By adopting a well analysed scenario, this alarm message sequence provided a degree of realism lacking in previous tests. The scenario was selected to ensure that any effects due to previous training and experience were approximately the same for all subjects.

Experimental conditions

Presentation modes

SELF-PACED
In order to determine subjects' preferred rate for reading messages, each new message was presented only after the current message had been categorized, enabling the subject to control the presentation rate.

EXTERNALLY-PACED

In order to determine subjects' maximum rate for reading messages, the messages were presented at increasing rates. Messages were presented in blocks of trials and within each block a fixed rate of presentation was used. Several blocks of trials were presented at steadily increasing message rates.

The general knowledge test was presented in the self-paced mode only. The verbal reasoning test and the alarm categorization test were presented in both the self and externally-paced modes. The alarm scenario categorization test was presented at the highest externally-paced presentation rate only.

Screen modes

Two modes of screen presentation were used in the experiment.

SCROLLED

All four tests were presented in this mode. Single line messages appeared on the screen until the display was full. After that, as a new message appeared at the bottom of the screen, the existing messages scrolled upwards one position, so that the top message disappeared from the top of the display. However, in the externally-paced trials if subjects categorized messages more slowly than the presentation rate, eventually the current message would move to the top of the screen. In this case it remained on the screen until it had been categorized and messages in the second position scrolled off the display. Any messages which scrolled off the screen without being categorized were defined as misses.

PAGED

The externally-paced verbal reasoning and the alarm categorization tests were also presented in a paged mode. This screen presentation mode represented more closely the way alarms are currently presented in AGR control rooms.

When the screen had filled with messages, any new messages were stored on a subsequent page which was not presented until the subject had dealt with the last remaining uncategorized message on the current page. The number of 'stored' pages was shown as a page number in the form 'PAGE 1 OF 3'.

Presentation rates

The alarm message presentation rates were determined on the basis of subject performance in pilot trials. In order to observe subjects' performance degradation at increasing message presentation rates, messages in the first block of externally-paced trials were presented at the slowest rate, then increased with each subsequent block of trials. Five blocks were used with presentation rates: 30, 75, 100, 110 and 120 messages per minute. Because it was only feasible to have a single alarm scenario categorization test, the highest message presentation rate (120 per minute) was chosen to ensure that noticeable subject performance degradation occurred.

Experimental facilities

A computer test system was implemented on a Unix workstation with a large high resolution colour graphics VDU (19 inch, 1024 × 768 pixels). The tests were designed so that subjects used only three clearly labelled buttons. All the tests and some of the post-test analysis were fully automated so as to ensure test repeatability and to provide a rapid analysis and summary of results.

In order to cause the minimum of disruption to station activities at the three AGR power stations, a portable human factors laboratory was transported to each of the three stations in turn. The portable laboratory provided sufficient space for the computerized test system, subject and two experimenters.

Subjects

In order to improve experimental validity, it was important to use subjects who were experienced control room operators. Subjects were operators, more properly termed 'desk engineers', from the three AGR power stations. These operators were sufficiently motivated to get fully involved in the experiment as they were familiar with full-scope simulator training, which in their view was a similar activity. In total 40 male operators volunteered to take part in the experiment during their normal shift duties. Activity at the stations and the availability of relief staff affected the number of volunteers at each station. For the total subject population, the subjects' ages ranged from 25 to 47 years and the mean age was 37 years.

Experimental design

The experiment used a simple repeated measures design, with every subject attempting all of the tests in the same order. The fixed order of tests enabled the experiment to progress in a way that was more natural to the subjects. Additionally, the repeated measures design made the most efficient use of the limited number of experienced operators and coped with the indeterminate number of volunteer subjects at each power station. A more detailed description of the experimental design and the complete experimental procedure, which lasted approximately 50 minutes, are given in Hollywell (1990).

Measures

Objective measures

For each test in the experiment the following data were recorded on-line by the test system for each subject:

- the category and duration of each displayed message;
- the response category and response time for each categorized message;
- the position (line number) of each categorized message when it was categorized;
- missed messages.

For each test in the experiment the following subject performance measures were produced on-line by the test system for each subject:

- mean accuracy of categorized messages (%);
- mean response time for categorized messages (seconds);
- number of categorized messages;
- number of missed messages;
- total number of messages displayed.

Subjective measures

Subjective data about the attitudes and feelings of subjects provides a valuable supplement to the more objective experimental measures. During alarm categorization tests it was considered useful to assess a subject's feelings after each trial block by means of a simple questionnaire employing bipolar rating scales. A slightly extended questionnaire was administered at the end of the experiment during debriefing to assess the subject's wider feelings about the experiment and their participation.

Discussion of results

Derived measures

The results of each test carried out during the experiment were considered in terms of the four derived measures.

- Mean accuracy (%) – mean value of subjects' mean accuracies of categorized messages in a test.
- Mean response time (seconds) – mean value of subjects' mean response times for categorized messages in a test.
- Mean ratio – mean value of subjects' miss ratio in a test, where miss ratio was calculated by dividing the total number of messages displayed during the test by the number of messages categorized. (This was considered a more precise measure of miss rate than a simple count of missed messages,

which would not take account of any uncategorized messages remaining on the screen at the end of the trial).
• Subjective ratings (%) – mean value of subjects' subjective ratings following a test.

Results of tests

The results of the verbal reasoning and alarm categorization tests are summarized in Table 3.1 below.

Verbal reasoning tests	Alarm categorization tests
Subjects maintained a mean accuracy of categorization in the region of 80% (ranging between 67.5% and 87.5%) in all externally-paced, scrolled screen tests irrespective of increasing message presentation rates.	Subjects maintained a mean accuracy for categorizing alarms in the region of 80% (ranging between 73.8% and 83.6%) in all externally-paced, scrolled screen tests irrespective of increasing alarm presentation rates.
Subjects' mean response time for categorization was in the region of 3.5 to 5.1 seconds (i.e. about 14 messages per minute) in all externally-paced, scrolled screen tests irrespective of increasing message presentation rates.	Subjects' mean response time for categorization was in the region of 1.5 to 2.6 seconds (i.e. about 30 alarms per minute) in all externally-paced, scrolled screen tests irrespective of increasing alarm presentation rates.
In the self-paced, scrolled screen test, the mean response time for categorization was significantly higher than that in the slowest externally-paced presentation rate by approximately 1.0 seconds.	In the self-paced, scrolled screen test, the mean response time for categorizing alarms was 4.2 seconds (ranging between 3.5 and 4.7 seconds), corresponding to a preferred alarm handling rate of 13 to 17 messages per minute.

[1] Performance was highly consistent among the three power stations. Negligible differences in mean accuracy and mean response time were observed.
[2] In the externally-paced tests, there was little difference in the mean accuracy of categorization or in mean response time between scrolled or paged presentation modes.
[3] The main degradation in performance was seen in the rate at which subjects missed messages/alarms. The missed message/alarm rate steadily increased with increasing message/alarm presentation rate.

Subjective ratings

• Subjects' subjective ratings indicated that they felt that they had worked harder, performed poorer, were more stressed and less relaxed in the externally-paced tests. The exception to this was when the externally-paced tests were given in the paged screen mode.

• Subjects did not find that the experiment became progressively fatiguing or
  boring.

Real-world situation

It should be noted that the experiment was a highly idealized representation
of the operator's task during a plant transient and factors in a real situation
could well affect the rate at which an operator can read alarm text messages.
In reality factors that might lead to faster performance than observed in
these tests are:

• the possibility for scanning down the alarm list;
• the alarms will often be in a logical, transient-dependent sequence not in
  random order;
• the operator may be allowed by the system to 'categorize' a whole page of
  alarms at the same time;
• the display system will not have messages disappearing off the top of the
  screen or continuously scrolling, which was probably a slight source of
  distraction in the experiment; and
• the operator has access to a rich variety of contextual information.

Factors in the real-world that might lead to slower performance than observed in these tests are:

• there were no extraneous distractions in the laboratory situation;
• alarm handling is only one of many tasks the operator must perform during
  a plant incident;
• the laboratory high resolution screen was of higher optical quality and
  produced better fonts than is typical of a control room VDU; and
• there is need for more consideration of the event itself in a real situation;
  this task required responses to single alarms only.

Nevertheless, the results clearly imply that if an operator is ever required to
read and understand more than 30 alarm messages per minute, then he will
either miss alarms or will create a backlog of unread alarms. Which of these
alternatives occur may well be determined by the mode of alarm message
presentation. In order to support an operator during periods of high alarm
arrival rate, consideration should be given to the implementation of automatic processing of alarm information, to reduce its volume and increase its
relevance to the operator. Close attention should also be given to the ergonomics of the alarm presentation.

Conclusions

The results obtained from this experiment were extremely consistent among
the three power stations. Because of this consistency and the relatively large

subject sample size (40 out of a total population of approximately 75), the following conclusions should therefore be highly representative of all operators' behaviour on these stations.

- The maximum rate at which operators can read alarm messages was in the order of 30 messages per minute when they had no other tasks to perform.
- Operators' subjectively preferred rates for reading alarm messages was approximately 15 messages per minute when they had no other tasks to perform.
- The nature of the operators' performance degradation observed over a range of increasing alarm presentation rates was not in accuracy or in response time; both of which remained remarkably constant. Degradation in performance was manifested in the number of missed messages, which rapidly increased as the alarm presentation rate exceeded the operators' maximum response capacity.
- The subjective data obtained from the questionnaires supported the above findings. It also confirmed that operators enjoyed the experiment.
- The potential mismatch between alarm presentation rates and reading performance indicates that consideration should be given to the implementation of automatic processing of alarm information, together with the close attention to the ergonomics of the design of alarm list systems.

Acknowledgements

The research described in this chapter was conducted in part fulfilment of P.D. Hollywell's MSc (Ergonomics), University of London, and with funding from Nuclear Electric, Power Gen and National Power. The study was undertaken whilst both authors were working at National Power Technical and Environmental Centre (NPTEC), Leatherhead. The authors would like to acknowledge the invaluable assistance given by the other members of the NPTEC human factors team in support of the experiment. The research is published with the permission of both National Power and Nuclear Electric. Please note that the views expressed in this chapter are solely those of the authors.

References

Baddeley, A.D., 1968, A three minute reasoning test based on grammatical transformation, *Psychonomic Science*, **10** (10), 341–2.

Baker, S.M. and Marshall, E.C., 1986, 'Evaluating the man–machine interface – the search for data', presentation at The 6th European Annual Conference on Human Decision Making and Manual Control, Cardiff, 2–4 June.

Barber, P.J., 1988, *Applied cognitive psychology: an information processing framework*, London: Methuen.

Danchak, M.M., 1988, Alarm messages in process control, *InTech (USA)*, **35** (5), May, 43–7.

Dillon, A., McKnight, C. and Richardson, J., 1988, Reading from paper versus reading from screen, *The Computer Journal*, **31** (5), 457–64.

EPRI, 1988, An evaluation of alternative power plant alarm presentations, *EPRI NP-5693Ps*, Vols. 1 and 2, Palo Alto, USA.

Fujita, Y., 1989, Improved annunciator system for Japanese pressurized water reactors, *Nuclear Safety*, **30** (2), 209–21.

Gould, J.D., Alfaro, L., Finn, R., Haupt, B. and Minuto, A., 1987, Reading from CRT displays can be as fast as reading from paper, *Human Factors*, **29** (5), 497–517.

Halstead-Nussloch, R. and Granda, R.E., 1984, Message-based screen interfaces: the effects of presentation rates and task complexity on operator performance, *Proceedings of the Human Factors Society 28th Annual Meeting*, pp. 740–4.

Hollywell, P.D., 1990, An experimental investigation into the rate at which process control operators can read alarm text messages: a cognitive engineering approach, MSc (Ergonomics) Project Report, Faculty of Engineering, University of London.

Jackson, A.R.G., 1988, 'The use of operator surveys by the CEGB to evaluate nuclear control room design and initiatives in the design of alarm systems and control room procedures', presentation at the IEEE Fourth Conference on Human Factors and Power Plants, Monterey, California, June 5–9.

Kemeny, J.G., 1979, Report of the President's commission on the accident at Three Mile Island, Pergamon Press.

Marshall, E.C., 1990, 'The national power man–machine interface prototyping and evaluation facility', presentation at the enlarged Halden programme group meeting, Bolkesjo, Norway, February.

Milne, R., 1984, Mistakes that mirrored Three Mile Island, *New Scientist*, 22 November.

Oborne, D.J. and Holton, D., 1988, Reading from screen versus paper: there is no difference, *International Journal of Man–Machine Studies*, Vol. 28, 1–9.

Reiersen, C.S., Marshall, E.C. and Baker, S.M., 1987, A comparison of operator performance when using either an advanced computer-based alarm system or a conventional annunciator panel, OECD Halden Reactor Project, *HPR-331*, Halden, Norway.

Sheridan, T.B., 1981, Understanding human error and aiding human diagnostic behaviour in nuclear power plants, in Rasmussen, J. and Rouse, W.B. (Eds.), *Human Detection and Diagnosis of System Failures*, New York: Plenum Press.

Singleton, W.T., 1985, Ergonomics and its application to Sizewell 'B', in Weaver, D.R. and Walker, J. (Eds), *The Pressurized Water Reactor and the United Kingdom*, Birmingham: University Press.

Singleton, W.T., 1989, *The Mind at Work*, Cambridge University Press.

4

Testing risk homeostasis theory in a simulated process control task: implications for alarm reduction strategies

Thomas W. Hoyes and Neville A. Stanton

Introduction

Ergonomists, particularly in the field of human reliability assessment, focus on ways in which process control tasks may be made intrinsically safer with a view to reducing human error and, it is hoped, accident loss. Alarm-initiated activity has properly come under their scrutiny. The rationale has been that human error can be reduced by reducing the complexity, the speed of presentation, and number of alarms. Yet the efficacy of such strategies presupposes that operators will not change their behaviour such as to negate any environmental benefit made by improvements to the human factors of alarms.

The theory of risk homeostasis, or RHT (Wilde, 1982a, 1982b, 1988, 1989) is one of several models that stress the importance of human factors in interventions aimed at improving the level of environmental risk. RHT holds that it is the target level of risk, rather than the absolute level of environmental risk, that determines accident loss (which can be thought of as being more or less equivalent to actual risk). RHT therefore posits a population-level closed loop process in which target and actual risk are compared. Wilde defines this 'target' as the level of risk that the individual deems acceptable. It is made up of four 'utilities': the perceived costs of relatively cautious behaviour, the perceived benefits of relatively cautious behaviour, the perceived costs of relatively dangerous behaviour, and the perceived benefits of relatively dangerous behaviour. If one or more of these utilities change, a corresponding change in target risk can be expected to follow. Risk homeostasis, then, is not about risk taking for its own sake, but rather presents a picture of risk-taking behaviour built on the concept of utility. (We use the preposition *on* rather than *with* to imply that any falsification of the role of utility in determining risk-taking behaviour in response to a change in the level of intrinsic risk, would be bound to leave the theory in a state of conceptual ruin.)

How might one evaluate the claims made by the proponents of RHT? Four approaches have so far characterized the debate. One has been the construction of theoretical/cognitive and mathematical modelling (O'Neill, 1977). This involves predicting behaviour from utility terms. O'Neill introduces in his model the notion of a negative utility for the utility of accidents. The difficulty with modelling is that it has so far proved impossible to derive from it very much in the way of testable hypotheses. The models are not only unverified, they would seem to be unverifiable.

The second approach is to examine accident loss statistics before and after an intervention. Perhaps the best example here, and certainly one that has attracted a great deal of attention, is that of compulsory seat-belt wearing (Adams, 1985; Bohlin, 1967, 1977; Bohlin and Aasberg, 1976; Chodkiewicz and Dubarry, 1977; Foldvary and Lane, 1974; Hurst, 1979; Lund, 1981). Next, there is the quasi-experimental study (Lund and Zador, 1984; Smith and Lovegrove, 1983). This involves taking measures of specific driver behaviours, such as speed, headway and so on, before and after some local intervention aimed at improving intrinsic safety. There are several difficulties with this. Firstly, just as in any quasi-experimental methodology, it can be difficult to disentangle cause from effect. Secondly, since Wilde (1988) is clear that RHT does not predict the particular behavioural pathway through which the effect will manifest itself, if such studies fail to confirm RHT predictions the proponents of that theory can point to the dependent measures that were sampled and argue that RHT would not have predicted changes. As an interesting side note, it is perhaps worth stating that in doing this, risk homeostasis theorists have not always played with a straight bat. In cases where speed changes after an intervention, this is held to support the theory; yet in cases where no speed changes are observed, they point out that the particular behavioural pathways are not predicted by RHT to carry the effect (McKenna, 1986 has a more detailed discussion of this).

A third major difficulty with the quasi-experimental study, as highlighted by Hoyes, Dorn *et al.* (1992), is that, in common with the analysis of accident loss statistics, it only addresses half of the RHT question – the consequences of a change in intrinsic risk. Whether individuals are characterized by a target level of risk, and whether this target can be shifted via changes in relevant utilities, are questions that the quasi-experimental study can never answer.

The fourth methodology associated with RHT is the simulation approach. Many attempts have been made to understand risk homeostasis theory in simulated risk-taking environments. Examples of this approach are provided by Mittenecker (1962); Näätänen and Summala (1975); Veling (1984); Wilde, Claxton-Oldfield *et al.* (1985); Tränkle and Gelau (1992). Hoyes, Dorn *et al.* (1992) and Hoyes and Glendon (1993) argue that all of these early simulations are flawed in that they rely for their validity on the generalization from non-physical to physical risk-taking, and that this assumption is unwarranted. Moreover, they point out that the awarding of points as a substitute for the real utility of a risk-taking experience is both conceptually inappropriate and laden with demand characteristics. Thus, the simulated examination of risk homeostasis theory undertaken by Hoyes, Dorn *et al.* (1992) involved some attempt to simulated physical risk.

The question of whether RHT can be examined in the laboratory in simulated conditions is very much open to question. Hoyes (1992) reports studies in which a simulator – the Aston driving simulator – has been evaluated using verbal protocol analysis, factor analysis, focus groups, and questionnaires. The evaluation of process control simulators for testing RHT predictions that might have some bearing on alarm design has yet to be undertaken.

Although it is difficult to envisage ways in which the utility question can ever be answered through a quasi-experimental approach, Hoyes, Dorn *et al.* (1992) suggest that utility can be investigated in a laboratory environment. Using a validated driving simulator (the Aston driving simulator), the experimenters attempted to operationalize the factors of intrinsic risk and utility in a controlled experiment. Intrinsic risk had three levels. In one condition, participants were informed that the simulator was fitted with an advanced braking system that would enable the vehicle to stop very quickly, if necessary. The information here was false and the braking system was in fact 'standard'. In another condition, the same information was given about the braking system, but this time the information was true in that pulling-up distance was reduced. Finally, a control condition was included in which the standard braking system was used, with no information given about braking efficiency. Having operationalized intrinsic risk in this way, the researchers turned their attention on how they might study utility. Although the researchers realized that personal utility could not easily be measured as a dependent variable, they believed that it could be manipulated as an independent one, so long as the variable in question objectively differentiated between benefits and costs. This was achieved by a time/distance design. In one condition, the experimental session was said to last for a period of ten minutes, whilst in another condition, the session was said to be over only when a distance of 7.2 miles had been covered. The experimenters argued that risk-taking behaviour had, objectively, a greater utility on the *distance* condition, since risk-taking could be expected to reduce in time what was quite an arduous task. On the *time* condition, the experimenters argued that risk-taking behaviour had relatively little utility. After the study was complete, the experimenters argued that, since the factor of utility had produced very large main effects, its operationalization must have been successful. The findings of the study could be summarized by saying that intrinsic risk affected some behaviours, most notably speed variability, whilst utility produced large effects across a wide range of specific driver behaviours. What was interesting was that the interaction of intrinsic risk and utility, predicted by RHT, was not observed, thus questioning the role of utility in determining changes in behaviour in response to a change in intrinsic risk.

It is important, however, to recall that in Wilde's (1982a, 1982b, 1988) model of risk homeostasis theory, it is suggested that the mechanism by which an equilibrium state of accident loss is said to take place involves three separate behavioural choices (Wilde, 1988, proposition 2). When a change is made to the level of environmental risk, the risk-taker may respond first by 'behavioural adjustments within the risk-taking environment'. In a road traffic environment, this may involve driving faster or slower, overtaking less frequently, reducing the marginal temporal leeway at which an overtake will be attempted, increasing or decreasing attention, and so on. A second route to the achievement of homeostasis is what one might term 'mode migration' – changing from one form of transport to another. For example, a motorcyclist may decide, in the light of inclement weather, to take a train into work rather than risk collision by using his or her motorcycle. Finally, if the level of target risk and the level of actual risk cannot be reconciled, either within the risk-taking environment, or through changing from one mode of transport to another, the individual may elect to stay at home and not to undertake any journey. This possibility, for the purposes of this paper, will be referred to as 'avoidance'.

So, the achievement of risk homeostasis can, according to its originator, be brought about in three ways. These can be labelled *behavioural adjustments within the environment*, *mode migration*, and *avoidance*. Out of this comes a realization that all of the above attempts to examine RHT in simulated environments have, in fact, looked only at one possible pathway to homeostasis: behavioural adjustments *within* the environment. Interesting though this question is, it would appear potentially to answer only one third of the risk homeostasis question.

The study reported here then has several aims. First, it seeks to examine once again the possible interaction between utility and intrinsic risk. But rather than investigating the interaction within the risk-taking environment, it seeks to examine evidence for it through the pathway of avoidance. Second, the study is concerned with risk-taking behaviour in an environment in which intrinsic risk is so great that on the high risk condition, accident loss is inevitable. In the Hoyes, Dorn *et al.* (1992) study, their high risk condition did not inevitably lead to high levels of accident loss, but only did so in interaction with specific behavioural decisions, such as electing to carry out a high-risk overtake. Third, the study aims to extend RHT research beyond the road-traffic environment and into a more general physical risk-taking environment – a simulated alarm-handling task.

Method

Participants

Forty five participants took part in this study. All were first year psychology students from The University of Aston. All were aged between 18 and 39 years.

Equipment

Forty five Macintosh IIsi microcomputers were used in the simulated alarm handling task. Each had a program simulating a control room. This program was originally coded in Supercard 1.5, but ran as a stand-alone application.

Design and procedure

Two tasks were performed – a primary and a secondary task.

The primary task

This was a matching/categorization task. To the right of the screen a number of alarms were presented. To the left of the screen four target boxes and a non-target box were shown (see Figure 4.1). The participants' task was to categorize the top, highlighted alarm to the right of the screen as either one of the targets, or as a non-target. This was achieved by moving a cursor by a mouse control to the appropriate selection box and clicking the mouse control.

The secondary task

In addition to the primary task, participants were asked to carry out a secondary spatial decision task. For this, a stick-figure was presented to the left of the screen holding an object in one hand. To the right of the screen a second figure was shown, but on a different rotation from the first. The task was to decide whether the figures matched or did not match. The direction buttons of the Macintosh were labelled 's' and 'd' (same and different). After pressing one of these buttons a new rotation was presented, and so on. The secondary task performance could therefore be measured in terms of the number of attempts and in terms of task accuracy.

General

In addition to the screen instructions, participants were told that the alarm to which they should refer at all times was the top highlighted alarm. This information was given three times: before the primary task practice, before the combined task practice, and before the combined task-proper. Participants were also told that the primary task should, at all times, be given priority.

Participants were informed that a prize of £5 would be given to the best score in each condition, though they were not informed what criteria would be applied to determine 'best' performance. The reason for not disclosing the way in which the performance on the relative measures was converted into

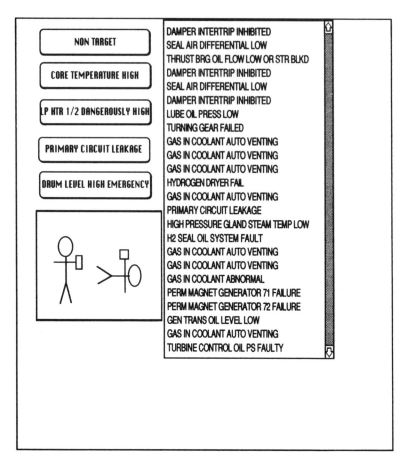

Figure 4.1   The screen display design during the experiment.

a single measure for comparison was that no objectively correct course of action could be said to exist at the outset. The pathway of avoidance was not included in this prize. The reason for offering a prize at all was to maximize the probability that individuals would be characterized by a target level of risk.

The pathway of notional avoidance was included in the form of a question that appeared to participants after the simulation exercise was complete. This asked participants how likely they would be to remain in the environment simulated for three different salary levels (£6000 p.a., £15 000 p.a., and £28 000 p.a.). Likelihood scores ranging from 1 (described as *not very likely*) to 10 (described as *very likely*) were recorded.

A practice was given for the primary task, the secondary task, and the combination of primary and secondary tasks. During the experimental session two experimenters remained to answer questions from the participants.

The experimental design

This was a two-factor design, each factor reflecting environmental risk. The primary risk factor was temporal probability of target, which can be thought of as presentation rate. It had three levels: one alarm per 1, 4 and 8 seconds. The second factor was ratio probability of target. This too had three levels: 2, 6 and 10 per cent of alarms being targets. Presented alarms were categorized as: correct target, correct non-target, incorrect target, incorrect non-target, and missed alarm (a missed alarm was one that scrolled off screen without being processed by the participant).

The simulated alarm handling task was produced on a Macintosh IIsi microcomputer. Figure 4.1 shows the screen design which appeared.

Results

The effect of environmental risk on avoidance

Of primary interest in this study was the question of whether an interaction would be found between environmental risk and utility on the measure of avoidance. Surprisingly, environmental risk did not produce an effect of avoidance by itself ($F[2, 36] = 1.227$, NS) (F is the probability statistic used in the analysis of variance (ANOVA) and NS is not significant). Figure 4.2 shows the effect of environmental risk on the measure of avoidance:

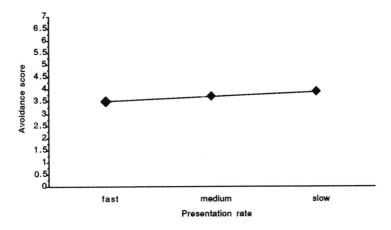

Figure 4.2 The effect of presentation rate (environmental risk) on the measure of avoidance.

Utility, by contrast, did produce an extremely large effect of avoidance ($F[2, 36] = 91.4$, $p < 0.0001$). However, the hypothesized interaction between environmental risk and utility was, just as in the Hoyes, Dorn *et al.* (1992) study,

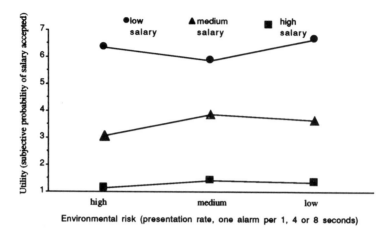

Figure 4.3   Environmental risk, utility and avoidance (low scores represent high likelihood of avoidance).

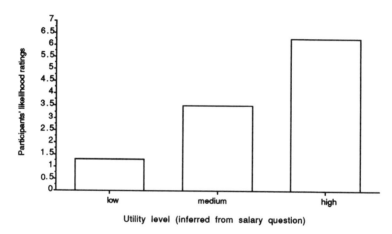

Figure 4.4   The effect of utility (salary for simulated job) and participant rating of likelihood of remaining in the environment.

not significant ($F[4, 72] < 1$, NS). It would seem then, so far as findings from this experiment goes, that environmental risk does not in any sense determine a participant's likelihood of removing him/herself from the environment. Figure 4.3 shows the relationship between environmental risk, the utility of remaining in the environment and avoidance. Figure 4.4 shows avoidance and the effect of utility. In interpreting Figure 4.4 it should be remembered that high ratings on the likelihood measure represent low judged probabilities of avoidance. In other words, participants characterized by high scores would be unlikely to remove themselves from the environment on the relevant utility condition.

Table 4.1 *Correlation coefficients between estimate of accident loss and notional avoidance (none is significant)*

Utility Level (Salary decision)			Accident loss measure
A	B	C	
−.072	−.209	−.081	Errors including missed targets
0.237	0.108	0.170	Errors excluding missed targets
−.117	−.234	−.107	Number of targets missed
−.121	−.238	−.117	Proportion of targets missed

Before rejecting altogether the possibility that a relationship might exist between environmental risk and utility on the measure of avoidance, a series of correlations were carried out, *post hoc*, between avoidance scores on each of the three levels of utility and four estimates of accident loss: errors including missed targets, errors excluding missed targets, the absolute number of targets missed, and the proportion of targets missed to incoming targets. This would appear to indicate that however accident loss is measured, and whatever utility level is examined, there is no relationship between errors (the indirect measure of accident loss) and rated probability of avoidance. The results of these *post hoc* tests are shown in Table 4.1.

Primary task performance

One might at this stage ask whether the above finding can be explained in terms of an operational negation of environmental risk. Was it the case that all participants were equally safe within the simulated environment?

To answer this question, a measure corresponding to probable accident loss is required. Three possibilities exist. First, the proportion of errors (incorrect hits plus incorrect misses divided by the total of incoming alarms) could be examined to provide an indication of likely accident loss. The problem with this measure is that it does not reflect errors of omission (missed alarms that scrolled off the screen without being acknowledged as either target or non-target). For this reason a second possibility for measuring likely accident loss would be to add missed targets to incorrect hits plus incorrect misses, dividing this total by total incoming alarms. Finally the proportion of missed alarms to total incoming alarms could be examined.

On the first measure of accident loss, the measure excluding missed alarms from the error criterion, the ratio probability factor did have an effect ($F[2, 36] = 3.274$, $p = 0.0494$). But the factor of temporal probability, deemed to be the main factor of environmental risk, showed no effect ($F[2, 36] = 2.316$, NS). The interaction between factors was not significant.

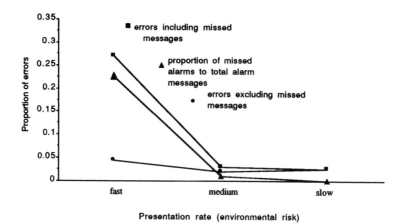

Figure 4.5   Presentation rate (environmental risk), proportion of errors and the three estimates of accident loss.

Interestingly, when one includes missed alarms in the error criterion, the significance of temporal and ratio probability are reversed. Ratio probability is now non-significant ($F[2, 36 = 2.614$, NS). Temporal probability moves from non-significance to $F[2, 36] = 12.902$, $p < 0.0001$. Again, there is no statistical interaction. Proportionately far more errors are recorded on the one-second presentation rate (the highest level of environmental risk) than on the four second and eight second rates.

When looking at the proportion of missed alarms there is again no ratio effect ($F[2, 36] = 1.505$, NS), but again there is a very large effect of temporal probability ($F[2, 36] = 10.214$, $p = 0.0003$). Once more there is no evidence for a statistical interaction of the two factors. Figure 4.5 shows the three estimates of accident loss in relation to presentation rate (environmental risk).

Secondary task performance

There are two ways of examining secondary task performance: in terms of the number of secondary tasks attempted or in terms of the error rate on those which were attempted. For the total secondary tasks attempted there was no effect of either ratio or temporal probability ($F[2, 36] = 1.431$, NS and < 1, NS respectively). For secondary task error rate the same was true with ratio probability giving ($F[2, 34] = 1.942$, NS), and temporal probability giving ($F[2, 34] = 1.862$, NS). In terms then of both error rates and total output, it seemed that participants were as accurate and as productive in the secondary task whatever the level of primary task demands made of them. The implications of these findings and a possible explanation of them is given in the discussion.

Discussion

This is the first simulation study to find evidence that is, on the surface at least, against risk homeostasis theory. In it, participants neither adjusted their behaviour within the environment such as to negate the adverse level of environmental risk, nor did they report that they would be more likely to leave the environment when it was more hazardous. On all relevant measures, when environmental risk was at its greatest, expected accident loss was at its worst. Therefore one could expect that improvements in environmental safety would result in commensurately reduced accident loss statistics.

The results of this experiment then point to one of two possibilities. Either this particular simulation was, for whatever reason, inappropriate as a tool for the investigation of RHT (these findings, in other words, have no practical relevance) or, and again for whatever reason, this experiment, though accurate in its relationship with the real environment it set out to model, represents a situation in which risk homeostasis does not usually occur.

Might it then have been the case that this study failed to uncover evidence for RHT simply because the environment simulated here is not characterized by a risk homeostasis process, at least in the short term? Although this is the first study to uncover what is, at first sight at least, evidence that could be interpreted as being against RHT, it is also the first study deliberately set up to prevent a complete negation of environmental safety behavioural change from within the simulated environment, is, one imagines, that such negation was not possible. (To put it bluntly, participants carried out either a *possible* task, or one of two *impossible* tasks.) This only leaves the measure of avoidance to be explained in which a negation of environmental risk behavioural change was possible by participants. This notional 'avoidance' pathway would be equivalent to participants saying, in effect, that whilst they could not maintain a constant level of accident loss within the environment, they could and would get out of that environment. This did not happen. On the measure of notional avoidance participants did not significantly react to the level of environmental risk, and when the correlation between the various measures of accident loss and avoidance scores was examined, these correlations too failed to reach significance. In other words, participants who had made a large number of errors were not significantly more likely to report that they would leave the environment than participants making fewer errors. To some extent this might be explained in terms of removal of feedback, for feedback of false negatives and false positives (active errors) was not given. Perhaps then participants were not aware that they were making errors. However, this explanation is inadequate in as much as feedback of passive errors was given. Missed targets were highlighted as they scrolled off the screen, making participants aware that a passive error had occurred. Even when the the error criterion includes passive errors (or, for that matter, consists entirely of passive errors as in the dependent measure of missed targets), the correlation between the error criterion and avoidance, across all three levels of utility, is not significant.

The conclusion from this would seem to be that where behavioural compensation within the environment is not possible, participants show a marked reluctance to engage in external compensation by removing themselves from the environment. Tentatively, then, one might conclude that where environmental safety standards fall to a level at which participants are unable to change their behaviour within the environment to negate the change (examples here might include particularly unfavourable weather conditions, snow and ice etc., or, as simulated here, increases of a certain magnitude in mental workload demands), the change in environmental risk will be matched by a commensurately large increase in accident loss. The other side of the coin, of course, would be where the level of environmental safety is such that accident loss inevitably characterizes the environment (the same examples as above applying), and environmental safety then changes such that the pre-existing level of accident loss is no longer inevitable, then the environmental improvement will lead to a decrease in accident loss. Perhaps then, in these circumstances at least, an engineering solution to the operational safety question would be possible.

Another interesting feature of this experiment is the complete failure of the secondary task to differentiate between any levels of the ratio probability factor or temporal probability factor. This was true both in terms of total output (ignoring accuracy) and in terms of accuracy (ignoring total output). This finding was initially difficult to reconcile with the design, since it would seem to imply that the attention demands made in the primary task do not affect performance, in any way, on the secondary task. It would seem, by extension, to indicate that an alarm presentation rate of one-per-second demands no more resources than an alarm presentation rate of one-per-eight-seconds. Even over the N of 45, there was no evidence even of a non-significant trend on any comparison.

What is in all probability the answer to this enigma was put by a participant to one of the experimenters sometime after the experiment (and at a time when re-analysis and a checking of raw data was actually taking place). It was suggested, from the participant's own experience, and from talking to others who took part in the study, that the higher levels of demand in the primary task were just too difficult. Rather than attempt to do the impossible, to cope with a presentation rate of, for example, one alarm every second, participants stopped even trying to do the primary task and put all of their efforts into the secondary task, which was, after all, considerably easier. This, it was suggested, allowed participants at least to salvage something from the study (there was, of course, the possibility of a prize to consider). In view of this, it would in all probability be inappropriate to consider further the meaning of the secondary task results. They were, it now seems, an artifact of the experimental design.

What is clear from this study, it must be remembered, is only that a constancy of actual risk was falsified. But RHT does not necessarily predict risk constancy, and in cases where the target level of risk of those individuals affected by the environment changes also, then a change in actual risk would in fact be predicted by the theory. So could it be argued that the target level of risk across the differing conditions of environmental risk was not constant? In answering this question it will be recalled that Wilde (1988) suggests that the target level of risk comes from four relevant utilities: costs and benefits of relatively cautious behaviour and costs and benefits of relatively risky behaviour. Perhaps then it could be argued that on the pathway of avoidance – leaving the environment – the costs of relatively cautious behaviour have risen, and stand at much higher levels than those that would be associated with behavioural adjustments within the environment. This being so, one might predict that the target level of risk would change such that individuals would be prepared to accept higher levels of accident loss, precisely what has happened here!

Before closing, it is interesting to note that, in a recent review of the literature, Hoyes and Baber (1993) and Hoyes (in press) suggest that RHT may be a difficult theory generally to apply to non-transport-related risk, and, in particular, to process control risk. In addition, they note that whereas RHT is built on the concept of time utility, process control tasks have little if any such utility. On the basis of this review, and on the basis of the study reported here, it would seem likely that interventions made to alarm environments will not be negated through the behavioural adjustments of the operators.

References

Adams, J.G.U., 1985, *Risk and Freedom: The record of road safety regulation*, London: Transport Publishing Projects.

Bohlin, N.I., 1967, A statistical analysis of 28 000 accident cases with emphasis on occupant restraint value, *Proceedings of the 11th STAPP conference*, SAE, New, York.

Bohlin, N.I., 1977, Fifteen years with the three point safety belt, *Proceedings of the 6th Conference of IAATM*, Melbourne, Australia.

Bohlin, N.I. and Aasberg, A., 1976, A review of seat belt crash performance in modern Volvo vehicles, *Proceedings of Seat Belt Seminar*, conducted by the Commonwealth Department of Transport, Melbourne, Australia.

Chodkiewicz, J.P. and Dubarry, B., 1977, 'Effects of mandatory seat belt wearing legislation in France,' presentation at The 6th Conference of IAATM, Melbourne, Australia.

Foldvary, L.A. and Lane, J.C., 1974, The effectiveness of compulsory wearing of seat belts in casualty reduction, *Accident Analysis and prevention*, 59–81.

Hoyes, T.W., 1992, Risk homeostasis theory in simulated environments, unpublished PhD thesis, The University of Aston in Birmingham.

Hoyes, T.W., Risk homeostasis theory – beyond transportational research, *Safety Science*, in press.

Hoyes, T.W. and Glendon, A.I., 1993, Risk homeostasis: issues for future research, *Safety Science*, 1ь, 19–33.
Hoyes, T.W. and Baber, C., 1993, Risk homeostasis in a non-transportational domain, in Lovesey, E.J. (Ed.) *Contemporary Ergonomics 1993*, pp. 178–183, London: Taylor & Francis.
Hoyes, T.W., Dorn, L. and Taylor, R.G., 1992, Risk homeostasis: the role of utility, in Lovesey, E.J. (Ed.) *Contemporary Ergonomics 1992*, London: pp. 139–44, London: Taylor & Francis.
Hurst, P.M., 1979, Compulsory seat belt use: further inferences, *Accident Analysis and Prevention*, 27–33.
Lund, H.W., 1981, Komentarer til 'The efficacy of seat belt legislation', *The Danish Council of Road Safety Research*.
Lund, A.K. and Zador, P., 1984, Mandatory seat belt use and driver risk taking, *Risk Analysis*, **4**, 41–53.
McKenna, F.P., 1986, Does risk homeostasis theory represent a serious threat to ergonomics? In Oborne, D.J. (Ed.) *Contemporary Ergonomics*, pp. 88–92, *Proceedings* from the *Annual Ergonomics Society's Conference*, held at The University of Durham.
Mittenecker, E., 1962, *Methoden und Ergebnisse der psychologischen Unfaallforschung*, Vienna, Austria: Deuticke.
Näätänen, R. and Summala, H., 1975, A simple method for simulating danger-related aspects of behaviour in hazardous activities, *Accident Analysis and Prevention*, **7**, 63–70.
O'Neill, B., 1977, A decision-theory model of danger compensation, Accident Analysis and Prevention, **9** (3), 157–65.
Smith, R.G. and Lovegrove, A., 1983, Danger compensation effects of stop signs at intersections, *Accident Analysis and Prevention*, **15** (2), 95–104.
Tränkle, U. and Gelau, C., 1992, Maximization of subjective expected utility or risk control? Experimental tests of risk homeostasis theory, *Ergonomics*, **35** (1), 7–23.
Veling, I.H., 1984, A laboratory test of the constant risk hypothesis, *Acta Psychologica*, **55**, 281–94.
Wilde, G.J.S., 1982a, The theory of risk homeostasis: implications for safety and health, *Risk Analysis*, **2** (4), 209–25.
Wilde, G.J.S., 1982b, Critical issues in risk homeostasis theory (response), *Risk Analysis*, **2** (4), 249–58.
Wilde, G.J.S., 1988, Risk homeostasis theory and traffic accidents: propositions, deductions and discussion of dissension in recent reactions, *Ergonomics (UK)*, **31** (4), 441–68.
Wilde, G.J.S., 1989, Accident countermeasures and behavioural compensation: the position of risk homeostasis theory, *Journal of Occup. Accidents*, **10** (4), 267–92.
Wilde, G.J.S., Claxton-Oldfield, Stephen, P. and Platenius, Peter, H., 1985, Risk homeostasis in an experimental context, in Evans, L. and Schwing, R.C. (Eds) *Human Behavior and Traffic Safety*, Plenum Publishing Corporation.

Part 2
Considerations of the
human operator

Considerations of the human operator

Neville Stanton

Chapter 5 (by David Woods) explains the complexities of dynamic fault management and describes how human reasoning may be supported through the provision of intelligent aids. David points out that dynamic fault management has very different characteristics to static fault management. Typically the dynamic situation is characterized by:

- time pressure;
- multiple and interleaved tasks;
- high consequences of failure;
- time varying data.

He presents the process of abductive reasoning through a generic dynamic fault management scenario to illustrate limitations of human performance. On the basis of the problems highlighted, David suggests how cognitive aids could, to some extent at least, assist abductive reasoning in the context of dynamic fault management. However, he cautions that under certain circumstances intelligent systems could undermine human performance.

Chapter 6 (by Neville Stanton) presents a literature review within the context of a model of human alarm handling. The model, developed from questionnaire and observational studies, distinguishes between routine incidents involving alarms and more critical incidents involving alarms. The notion of alarm initiated activities (AIA) is used to describe the collective stages of alarm handling. The activities are intended to represent the ensuing cognitive modes and their corresponding behaviours that are triggered as a direct result of the presence of alarms. The six main AIAs are identified as: observe, accept, analyse, investigate, correct and monitor. It is argued by the author that each of these activities may have different information requirements which need to be supported if the alarm handling task is to be successful.

Chapter 7 (by Harm Zwaga and Hettie Hoonhout) consider the impact of distributed control systems on human supervisory control tasks (including alarm handling). They suggest that the introduction of distributed control systems was based upon the misconceived idea of human supervisory control largely consisting of operation-by-exception. Through a detailed analysis of operators tasks and activities Harm and Hettie suggest that current methods of displaying alarm information in a range of control rooms appears to be less than optimum. The problems appear to be particularly acute in times of major disturbances, ironically this is when the alarm system would be of most use to the operator. Harm and Hettie suggest a range of strategies that could be employed to substantially reduce the alarm load without resorting to added complexity and costs to the process control system.

Cognitive demands and activities in dynamic fault management: abductive reasoning and disturbance management

David D. Woods

Introduction

The cognitive activities involved in dynamic fault management are more complex and intricate than simple alerts on the one hand, or troubleshooting a broken device which has been removed from service on the other hand. Fault diagnosis has a different character in dynamic situations because there is some underlying process (an engineered or physiological process which will be referred to as the monitored process) whose state changes over time. Faults disturb the monitored process, and diagnosis goes on in parallel with responses to maintain process integrity and to correct the underlying problem. These situations frequently involve time pressure, multiple interacting goals, high consequences of failure, a great deal of time varying data and multiple interleaved tasks (Woods, 1988). Typical examples of fields of practice where dynamic fault management occurs include:

1. flightdeck operations in commercial aviation (Abbott, 1990);
2. control of space systems (Woods, Potter *et al.*, 1991);
3. anaesthetic management under surgery (Moll van Charante, Cook *et al.*, 1993); and
4. terrestrial process control (Roth, Woods *et al.*, 1992).

My colleagues and I have been engaged in a series of studies of dynamic fault management in the above four domains including:

- empirical studies in the field (Sarter and Woods, 1993; Cook, Woods *et al.*, 1991);

- studies of putative support systems (Moll van Charante, Cook *et al.*, 1993; Woods, Potter *et al.*, 1991);
- simulation studies of cognitive activities (Roth, Woods *et al.*, 1992); and
- the design of new aiding strategies (Potter and Woods, 1991; Potter, Woods *et al.*, 1992; Woods, in press).

This chapter describes the cognitive demands and activities involved in dynamic fault management based on a synthesis of field experience and the results of studies across the above domains and issues.

The point of departure is the cascade of disturbances produced by a fault in a dynamic process. This cascade of disturbances is explored to highlight the cognitive demands of dynamic fault management, particularly the disturbance management cognitive task. Secondly the idea, current in artificial intelligence, that diagnosis is a form of abductive inference is extended for dynamic fault management and joint human–machine cognitive systems. The attempt to construct a cognitive systems model of the interacting factors at work in dynamic fault management illuminates a variety of pitfalls and a variety of new directions in the use of new technological possibilities to support human performance in this class of tasks.

The cognitive demands of dynamic fault management

The cascade of disturbances

A fault in a dynamic process produces a cascade of disturbances. What are the implications of this in terms of the cognitive demands and the cognitive strategies of people engaged in fault management? Exploring the consequences of the cascade of disturbances help us recognize and characterize the disturbance management cognitive task (Woods, 1988).

A fault disturbs the monitored process by triggering influences that produce a time dependent set of disturbances (i.e. abnormal conditions where actual process state deviates from the desired function for the relevant operating context). Thus, faults can be seen as a *source* of influences that act on the monitored process. The influences produced by faults are abnormal relative to desired process function and goals. Faults initiate a temporally evolving sequence of events and process behaviour by triggering a set of disturbances that grow and propagate through the monitored process if unchecked by control responses. I define a disturbance as an abnormal state of some functional portion of the monitored process where actual process state deviates from the desired state for the relevant operating context.

In the absence of countervailing control influences, the fault acts to produce a *cascade* of disturbances that unfold over time due to the development of the fault itself (e.g. a leak growing into a break) and due to functional and physical interconnections within the monitored process. Thus, a fault acts as a force that disturbs the monitored process via a set of influences that act on the monitored process, that is fault-produced or abnormal influences. Figure

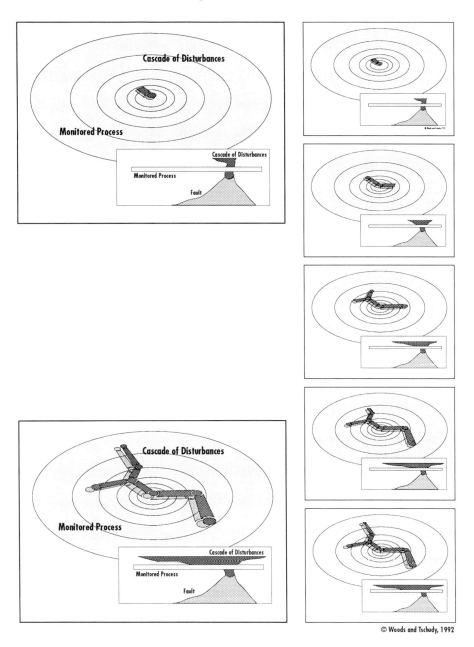

Figure 5.1   A fault produces a cascade of disturbances in a process.

5.1 illustrates a simple case of a fault producing a cascade of disturbances in a process. The cascade of disturbances is one forcing function that contributes to the difficulties in creating effective alarm systems. For example, the representation of the monitored process in alarms and displays can produce a cascade or avalanche of signals, sounds and messages when a fault produces

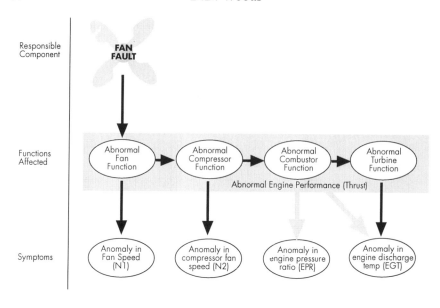

Figure 5.2   An aviation example of the cascade of disturbances that can follow from a fault (adapted from Abbott, 1990).

a cascade of disturbances (Potter and Woods, 1991; Reiersen, Marshall *et al.*, 1988).

Figure 5.2 provides an aviation illustration of the cascade of disturbances that can follow from a fault. The initiating fault is a failure in the fan sub-system of an aircraft engine. This fault directly produces an anomaly in one engine parameter, but the fault also disturbs compressor function which is reflected symptomatically in an anomaly in another engine parameter. The effect of the fault continues to propagate from the compressor to the combustor producing anomalies in two more engine parameters. Diagnosis involves understanding the temporal dynamics of the cascade of disturbances. For example in this case, the temporal progression is an important clue in understanding that the fault is in the fan subsystem and not in the compressor or the combustor. Note that, because of disturbance propagation, the same or a similar set of anomalies may eventually result from a fault in a different subsystem. A critical discriminating difference is the propagation path as the cascade of disturbances develops over time (Potter, Woods *et al.*, 1992).

Operational personnel and automatic systems act in response to the disturbances produced by a fault to maintain important goals, especially safety and system integrity and to correct the underlying problem. Manual and automatic actions interject control influences to counteract the abnormal influences produced by faults (Figure 5.3). These countervailing influences can act by

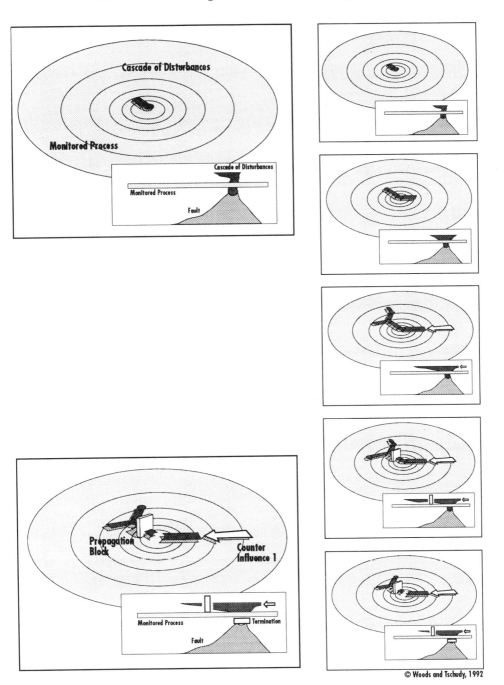

Figure 5.3 Operational personnel and automatic systems act in response to the disturbances produced by a fault to maintain important goals, especially safety and system integrity and to correct the underlying problem.

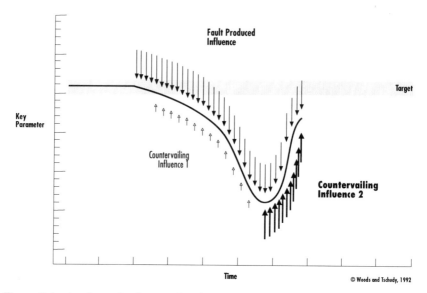

Figure 5.4   A schematized example taken from Woods, Pople et al. (1990) showing that the behaviour of the monitored process is a function of the combined effects of the fault-produced influences and the countervailing influences introduced by control systems and humans.

producing a force to counteract one of the influences produced by a fault (mitigate consequences), by producing a barrier against further propagation through the functional or physical structure of the monitored process, or by stopping the fault from producing any more abnormal influences (terminate abnormal influences).

The behaviour of the monitored process is a function of the combined effects of the fault-produced influences and the countervailing influences introduced by control systems and humans (Figure 5.4). In this case, the initial countervailing influence is unable to mitigate effectively the effects of the fault-produced abnormal influence on a key parameter. A second and stronger countervailing influence is then initiated which is able to overcome the fault-produced abnormal influence and redirect the parameter towards its desired range. The interaction of abnormal and control influences means that in some cases countervailing influences can act to mask the purely symptomatic indications that a fault is present or the purely symptomatic indications of the severity of a fault in the monitored process.

Decompensation

The masking effect of countervailing influences can lead to *decompensation* incidents. Decompensation incidents in managing highly automated processes are one kind of complication that can arise when automatic systems respond to compensate for abnormal influences generated by a fault. As the abnormal influences persist or grow over time, the capacity of the counter-influences to compensate becomes exhausted. At some point they fail to counteract and the system collapses or decompensates.

The presence of automatic counter-influences leads to a two phase signature. In phase 1 there is a gradual falling off from desired states over a period of time. Eventually, if the practitioner does not intervene in appropriate and timely ways, phase 2 occurs – a relatively rapid collapse when the capacity of the automatic systems is exceeded or exhausted. During the first phase of a decompensation incident, the gradual nature of the symptoms can make it difficult to distinguish a major challenge, partially compensated for, from a minor disturbance. This can lead to great surprise when the second phase occurs (some practitioners missing the signs associated with the first phase may think that the event began with the collapse (cf., exemplar case 1 in Cook and Woods, in press).

The critical difference between a major challenge and a minor disruption is not the symptoms by themselves but rather the force with which they must be resisted. Thus, lack of information about automatic system response can contribute to the failure to recognize the seriousness of the situation and the failure of the supervisory controller to act to invoke stronger counter-actions early enough to avoid the decompensation. Examples of decompensation incidents include the China Air incident in commercial aviation (National Transportation Safety Board, 1986) and a variety of incidents in anaesthesiology (Cook, Woods *et al.*, 1991). Anaesthesiology is particularly vulnerable to this type of incident because the automatic systems are the patient's internal physiological adjustments to chronic disease states. These are not well modelled and are difficult to see in operation. Decompensation is one kind of pattern that illustrates how incidents evolve as a function of the nature of the trouble itself and as a result of the responses taken to compensate for that trouble.

Incident evolution

Cognitive processes in situation assessment involve tracking and understanding the temporal evolution of the state of the monitored process as the interaction of multiple influences (Woods, Pople *et al.*, 1990; Roth, Woods *et al.*, 1992). Overall from an incident evolution point of view, faults initiate a temporal process where disturbance chains grow and spread, and then subside as control interventions are taken to limit the consequences and to break the chain of incident evolution (Woods, 1988; Gaba, Maxwell *et al.*, 1987). For example, Cook, Woods *et al.* (1991), in a study of critical incidents in anaesthesia, identified several different patterns of incident evolution. 'Acute' incidents present themselves all at once, while in 'going sour' incidents there is a slow degradation of the monitored process.

These situations are further complicated by the possibility of multiple faults each producing influences that act to disturb the monitored process. Interactions across individual fault-produced influences, as well as interactions with the countervailing influences introduced to respond to the trouble, can greatly complicate the diagnostic process of sorting out the relationship between perceived disturbances and possible underlying faults (Figure 5.5).

Another kind of diagnostic complication occurs when the disturbances produced by a fault quickly effect different parts of the monitored process (Figure 5.6). This can occur:

1. because the time constant of propagation is very fast across these parts;
2. because the fault produces multiple influences essentially simultaneously, each of which affects a different part of the monitored process; or
3. because the fault affects a function that is connected to many other process functions (e.g. a vital support system like electric power).

This kind of situation is a good test case for the evaluation of an alarm system prototype – does the alarms system simply become an uninformative 'Christmas tree' of lights, messages and sounds when a fault quickly disturbs a set of process functions?

The bottom line is that practitioners must be able to track patterns of disturbance propagation in order to be able to build and maintain a coherent situation assessment in dynamic fault management. One important question to ask in the design or evaluation of alarm systems then is – does a particular alarm system prototype support this cognitive function? (cf., Woods, Elm *et al.*, 1986 for one such concept). The ability to see disturbance chains grow and subside is important because new disturbances can be evidence that confirms previous hypotheses about the influence pattern acting on the monitored process, or new disturbances can be evidence that indicates that something different or new is going on (e.g. cues to revise what is an erroneous situation assessment; Woods, O'Brien *et al.*, 1987; De Keyser and Woods, 1990). Note how this aspect of fault management disturbs common notions about diagnosis. The behaviour of the monitored process is a joint function of the set of influences acting on it. In other words, an explanation of the behaviour of the process almost always (at least for any interesting case) contains multiple hypotheses (factors).

Symptoms, disturbances and model-based reasoning

It is commonly accepted that diagnosis is about the processes involved in linking faults to symptoms. But dynamic fault management situations reveal that faults are not *directly* linked to symptoms. Faults produce influences which disturb the function of the monitored process. Disturbances are abnormal conditions or malfunctions where the actual process state deviates from the desired function for the relevant operating context. The perceivable symptoms

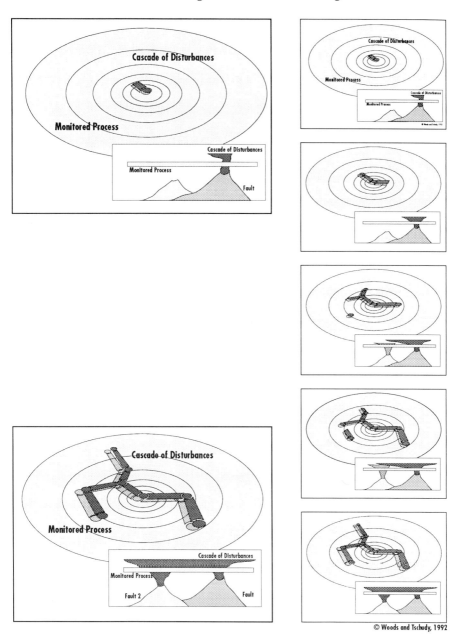

Figure 5.5   Multiple faults can produce interacting chains of disturbances which may be similar to the patterns produced by single faults (cf., Figure 5.1).

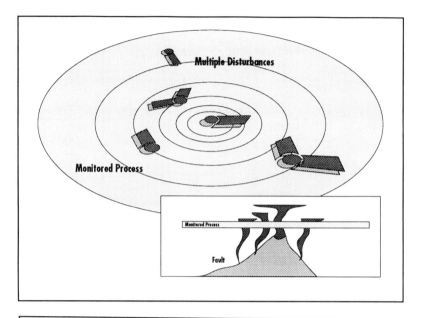

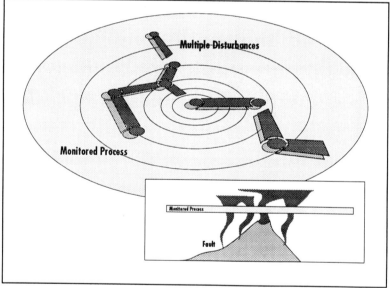

Figure 5.6   One kind of diagnostic complication occurs when the disturbances produced by a fault quickly effect different parts of the monitored process. Does the alarm system simply become an uninformative 'Christmas tree' of lights or messages or sounds when this situation occurs?

depend on the evidence that is available about disturbances relative to appropriate contrast cases, i.e. the sensor data and a model of proper function or expected behaviour (Rasmussen, 1986; Lind, 1991). Let us consider a simple example. For many processes (thermodynamic, physiological), one function that may recur is the concept of flow in a path to a destination. The concept of flow in a path is not the same thing as a flow measurement. The former refers to a state of a control volume (the path); the latter refers to a measurement at a particular place. A flow measurement can be evidence about the state of a flow process when combined with an implicit or explicit model of system structure and function. However, the relationship can be quite complicated depending on the circumstances.

For example (Woods, Roth *et al.*, 1987), typically one would expect a break in a system to produce reduced flow, i.e. the break will impair delivery of the material in question to the desired destination. However, in a pressurized system, a break downstream of the flow sensor that connects the high pressure piping to a low pressure region will produce a high flow measurement even though there is less material being delivered to the desired place. If there is a pressure sensor on the piping, one can see that the pressure difference between the pressure at the normal destination and the piping pressure will be much less than normal. In other words, following the break the pumping mechanism will be working against a lower pressure, thus for some kinds of physical systems delivering increased flow (but to an abnormal sink). The behaviour of the sensor depends on its location relative to the break. The behaviour of the sensor is a symptom; the concept of impaired delivery of material to the desired destination is the disturbance in the normal function of the process for that context (i.e. that it should be working in this phase of operation). Woods and Hollnagel (1987) and Woods, Elm *et al.* (1986) elaborate on disturbances in the function of a dynamic process.

The critical point is that the relationship between faults and symptoms fundamentally depends on how the fault disturbs the normal function of the process and on the details of the geometry of sensors.[1] This does not prevent a particular operator from attempting to understand the state of the monitored process in a purely symptomatic way, i.e. linking symptom to underlying fault. But it does point out why this is a very limited strategy if relied on exclusively in the case of dynamic fault management (Rasmussen, 1986; Davis, 1984; Woods, Roth *et al.*, 1987).

The above characteristics mean that diagnosis and fault management in dynamic applications have a very different cognitive character than the stereotype about diagnostic situations which is based on the exemplar of troubleshooting a broken device which has been removed from service.

---

[1] It may be that the dominance of medical diagnosis, or more properly, didactic exercises in internal medicine, as the prototype for diagnosis in general has encouraged this error. This is in part because good models of human physiology are lacking, because the diagnostician has extremely limited set of data measurements available about the state of the process, and because it seems that one can pretend that the faulted process is static.

Disturbance management

In dynamic fault management, the monitored process is not and usually cannot be removed from service. This means that the fault manager needs to try to continue to meet some of the goals of the monitored process (e.g. safety). The relative importance of different process goals may change as the incident evolves and some goals may need to be abandoned if they compete with more critical goals (Cook, Woods *et al.*, 1991). Thus, in dynamic, uncertain and dangerous domains, fault diagnosis occurs as part of a larger context where the expert practitioner must maintain system integrity by coping with the consequences of faults (i.e. disturbances) in parallel with untangling the causal chain that underlies these disturbances in order to take longer term corrective responses. The cognitive activities involved in managing the process to cope with the consequences of disturbances that go on in parallel with and interact with fault diagnosis I have called *disturbance management* (Woods, 1988). The disturbance management cognitive activity refers to the interaction between situation assessment and response management (response selection, adaptive planning and plan generation) that goes on in fault management in dynamic process applications. Disturbance management is not simply the aggressive treatment of the symptoms produced by some fault. It includes the continuing and purposeful search for the underlying fault(s) while at the same time struggling to preserve system viability and, especially, how these two lines are co-ordinated given time pressure, the possibility of bad outcomes and the need to revise provisional assessments as new evidence comes in over time (Roth, Woods *et al.*, 1992; Cook and Woods, in press).

Cook and Woods (in press) contains an abstracted protocol of one actual critical incident in anaesthesiology which involved some aspects of disturbance management. During this incident the physicians engaged successfully in disturbance management to cope with the consequences of a fault (itself the result of a breakdown in the person–machine system). The physicians were unable to identify the exact source of the incident until after the consequences of the fault had ended due to other factors. However, they were able to characterize the kind of disturbance present and to respond constructively (through the mitigation mode of response; see the section on modes of corrective response) in the face of time pressure without becoming fixated on pursuing what was the 'cause' of the trouble. In contrast, another study of anaesthesiologist cognitive activities, this time in simulated difficult cases, (Schwid and O'Donnell, 1992) found problems in disturbance management where about one third of the physicians undertreated a significant disturbance in patient physiology (hypotension) while they over-focused on diagnostic search for the source of the disturbance.

The disturbance management cognitive task is one case that raises questions about the relationship between diagnosis and action (Klein, Orasanu *et al.*, 1993). In disturbance management tasks, actions are likely to precede diagnosis and go on interwoven with diagnostic search activities. The effect that countervailing control influences have on disturbances generates information about the fault. For example, in nuclear power plants the rate of deterioration or stabilization in the face of interventions of increasing strength is one source of information about the size of piping breaks. This is just one example of how interventions are both diagnostic as well as therapeutic in fault management. Woods, Roth *et al.* (1987) and Schwid and O'Donnell (1992) report data that indicate that failure to attend to, or to integrate with, the diagnostic information that follows from an intervention is associated with failures to recover from erroneous situation assessments. In dynamic fault management, intervention precedes or is interwoven with diagnosis.

Coping with complexity: strategies in disturbance management

For the people caught up in an event-driven problem, the situation is fluid, unfolding. In other words, there is an open future for people in a problem. This means that there is a fundamental difference between the point of view of people in a problem and the point of view of an outside observer with knowledge of the outcome, i.e. the actual fault(s) present. The openness of the future for someone-in-the-problem interacts with the event-driven nature of these situations, that is new events may occur in the future to change the course of development of the incident. This means that what will happen in the future is not closed but rather open to various possibilities, and the person in the situation can and must act to influence which possibilities will come to pass. Note how this aspect of dynamic situations highlights the responsibility of the people-in-the-problem.

Human fault managers cope with the cognitive demands of fault management in a variety of ways. Diagnostic interventions are the rule; that is practitioners act on the monitored process, not necessarily to move it to a new state, but rather to generate information about the state of the process in terms of how it behaves in response to the intervention. For example, the practitioner will often take actions whose primary purpose is to check out or confirm a hypothesis about the source of the trouble. Another commonly observed tactic is to combine an information generation purpose and the 'need for safing' responses in the early stages of a developing incident by choosing an intervention where the same action both generates information about the state of the process and helps to maintain critical safety goals.

Another common strategy in disturbance management is, when confronted by an anomaly, to check on the control influences that are currently acting on the monitored process. Thus, we see disturbance managers asking what have I done recently to affect the monitored process? Did I do what I thought I did, e.g. anaesthesiologists will check did I give the wrong dose or the drug (Cook, Woods *et al.*, 1991; Cook and Woods, in press)? What are automatic systems doing to the monitored process or what have they been doing in the recent past? Has there been a failure in the automatic systems to act as demanded? This strategy occurs frequently in part:

- because of the need to detect erroneous actions or failures of systems to start or to respond as demanded;
- because of the possibility of erroneous actions that lead to activation of a different counter-influence than the one intended; and
- because of the possibility of co-ordination surprises where one agent (typically a human practitioner) is surprised by the behaviour of another team member (usually some automated system).

Modes of corrective responses

There are four basic modes of response that can be invoked to combat the abnormal influences produced by a fault in disturbance management. These are:

- mitigate consequences;
- break propagation paths;
- terminate source;
- clean-up after-effects.

Mitigate consequences

In this mode of response, one is just coping with the consequences of the fault by treating each threat to safety goals to preserve integrity of system. This response mode tends to occur at periods of fast deterioration in process health and when very little diagnostic information is available. For example, take the case of an unisolateable break. One cannot stop the break in the short run; rather one counteracts or neutralizes the disturbances produced by the fault, stabilizes the monitored process and then considers longer term response strategies to terminate the fault. Note that one does not have to know anything about the nature of the fault itself in this mode of response. Instead, one focuses on the threats to system integrity and safety.

Break propagation paths

This mode of response is also directed at the consequences of faults, counter-acting the influences produced by the fault with control influences. Again, one does not have to know anything about the nature of the fault itself in order to respond. In this mode, one acts to block the chain of disturbance propagation (usually an action blocks only some of the functional propagation paths). For example, isolation of a break stops the accumulation of material/energy in some undesired places, but the break is still an active influence. Therefore, loss of material/energy can continue and accumulation within some regions will still be going on.

Note that both of the these modes counteract the influences produced by faults and can be referred to as safing actions because the intent is to preserve safety goals and system integrity.

Terminate source

Responses in this mode stop the fault from generating abnormal influences that affect the monitored process. Thus, one no longer has to counteract with control influences as the fault is no longer a source of abnormal influences. One must know something about the source of the abnormal influences (the underlying fault) to use this mode of response. An example would be to stop the break (e.g. patch or equalize pressures) so that there is no more loss/accumulation of material/energy in undesired regions.

Clean-up after-effects

Some disturbances may persist even after fault termination. The after-effects are consequences of having had the fault present at all. This mode of response is directed at handling these effects after the fault has been terminated. The magnitude and nature of the after-effects are partially a function of the skill with which disturbance management was carried out (and partially a function of the faults themselves). For example, one still must deal with the effects of the break having occurred, i.e. the undesired accumulation of material/energy in abnormal regions and amounts.

In general the four modes can be seen as defining a sequence of corrective responses – safing responses occur first, followed by diagnostic search leading to termination responses and treating after-effects. This generalization is true but only in an aggregate way. For example, if strong diagnostic information is present and events are not moving too quickly, the first response may be to terminate the source. A quick and correct initial assessment of the source of the disturbances may occur, but a variety of factors can occur later during the response to the initial failure that lead to the need for new safing or diagnostic activities. For example, new additional faults can occur, there can be failures in carrying out the corrective responses (erroneous actions during manual execution or failures of components to respond as demanded), or the original diagnosis could turn out to be in error. Any of these events could lead to the need for additional safing activities, while new diagnostic work is going on to revise or refine the situation assessment.

Anomalies and expectations

In everyday usage, an anomaly is some kind of deviation from the common order or an exceptional condition. In other words, an anomaly represents a mismatch between actual state and some standard. Fault management is concerned with identifying what anomalies are present in the monitored process and developing a best explanation that, if true, would account for the observed pattern of findings (an abductive reasoning process).

To characterize a fault management system cognitively, one must specify the different categories of anomalies that the system can recognize, and the information processing activities needed to recognize these classes of events (Woods, Roth *et al.*, 1987; Roth, Woods *et al.*, 1992). One kind of anomaly has to do with departures of observed monitored process behaviour from the desired system function for a given context, i.e. the monitored process is not performing the way it was designed to perform. It could be that pressure is supposed to be within a certain range but that it is currently too low for a particular context (e.g. what is too low may vary with context such as shutdown versus full power operations in a process plant).

Another kind of anomaly has to do with process behaviour that deviates from practitioners' model of the situation. In this case, process behaviour deviates from someone's (operators') or something's (the intelligent system's) expectations about how the process will behave (Woods, Roth *et al.*, 1987). The agent's expectations are derived from some model of the state of the monitored process. Because we are focusing on dynamic processes, this model of expected behaviour refers to the influences acting on the process – influences resulting from manual actions; from automatic system activities, or from the effects of faults. Anomalous process behaviour that falls into this class we can call 'unexpected,' that is observed monitored process behaviour is unexpected with respect to model derived expectations, again for the particular context. Note that there may be other kinds of anomalies as well, for example, departures from plans.

Abnormal process behaviours may or may not be expected. For example, if you suddenly trip off a power generation system and there is some kind of coolant reservoir in the system, then level in that reservoir is going to drop (i.e. the rapid shutdown decreases the energy stored in the mass of liquid; so the volume occupied by the same mass decreases or shrinks). This decrease in level produces an alarm that level is outside limit values, but this event and this alarm always occur when the power generation system is rapidly shutdown. The level parameter is abnormal with respect to desired system function, and a corrective response should begin. However, the alarm is expected given the circumstances. The operator knows 'why' the alarm indication is present (it is an expected consequence of the influence of the rapid shutdown) and therefore this alarm does not interrupt or change his or her information processing activities. For example, the operator will not try to 'diagnose' an underlying fault. What would be unexpected would be the absence of this alarm or if the low level condition persisted longer than is expected given the influence of the event – the rapid shutdown.

Recognition of different kinds of anomalies should lead to different follow-up 'lines of reasoning' (Figure 5.7). Recognition of 'abnormal' process behaviour should lead to information processing about how to cope with the indicated disturbance, e.g. safing responses (in the example above, make-up systems are supposed to come on to restore indicated level to the desired range). This, in turn, leads to monitoring lines of reasoning – checking to see if coping responses have occurred as expected and whether they are having the desired effect. Thus, in the above example, the low level alarm should

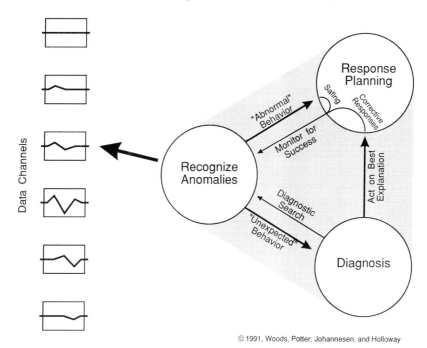

Figure 5.7 Schematic of anomaly-driven cognitive activities or lines of reasoning involved in the disturbance management cognitive task (from Woods, Potter et al., *1991).*

trigger a line of reasoning to evaluate what coping responses should be initiated to deal with the low level abnormality. In this case, an automatic make-up system should start up to resupply the reservoir. The low level alarm should trigger a line of reasoning to monitor that the automatic system came on properly and is restoring level to the desired range. Alternatively, recognition of an 'unexpected' process behaviour should lead to diagnostic information processing – a line of reasoning to generate possible explanations or diagnoses for the observed anomaly and knowledge-driven search to evaluate the adequacy of those possible explanations (Roth, Woods *et al.*, 1992).

These different lines of reasoning are intertwined and paced by changes and developments in the monitored process. Depending on the state of the fault management process (the mindset of the practitioners), the same incoming data may have very different implications. Consider the following examples. If the background situation assessment is 'normal system function,' then new incoming data about the state of the monitored process may be informative, in part, because they signal that conditions are moving into abnormal or emergency operations. If the background line of reasoning is 'trying to diagnose an unexpected finding,' then the new signals may be informative because they support or contra-indicate one or more hypotheses under consideration. Or, if the background line of reasoning is 'trying to diagnose an unexpected finding,' then new signals may be informative because they function as cues to generate more (or broaden the set of) candidate hypotheses that might explain the anomalous process behaviour. If the background line of reasoning is 'executing an action plan based on a diagnosis,' then the new data may be informative because they function as cues that the current working hypothesis may be wrong or incomplete since the monitored process is not responding to the interventions as would be expected based on the current working hypothesis.

The above characterization suggests that abductive reasoning is a part of the disturbance management cognitive activity. Abductive inference or inference to a best explanation is often invoked as a cognitive model of fault diagnosis in the artificial intelligence community (Josephson, 1991; Peng and Reggia, 1990). The basic template for abductive reasoning, as opposed to deductive or inductive reasoning, can be stated as follows (taken from Josephson, 1991; cf., Peirce, 1955):

1. D = collection of observations, givens, facts;
2. H explains D, that is, H, if true, would account for D;
3. no other available hypothesis explains D as well as H does;
4. therefore, H is probably true.

However, abductive models of diagnoses have not focused on dynamic fault management. How does abduction apply to diagnostic reasoning about the state of a dynamic process or device when new events can occur at indeterminate times and when there is a relatively large amount of data available about the state of that monitored process (Woods, 1988)? How does abductive reasoning function in the context of human–computer co-operative problem solving (Woods, Potter *et al.*, 1991; Woods, 1992)?

Abductive reasoning and fault management of dynamic processes

How do the reasoning steps in the basic template for abductive inference apply when one is reasoning about the state of a dynamic process given relatively large amounts of data about the state of that process or device?

STEP 1: D = COLLECTION OF OBSERVATIONS, GIVENS, FACTS
At this step, for data rich situations, the basic problem is defining what are the set of findings to be explained. The inherent variability of dynamic physical systems means that there are a large amount of changes in the monitored process (and the absence of changes) that could be relevant in principle. A sensitive mechanism is needed to extract from the continuous flow those changes or absences of change that are significant. A sensitive monitor

minimizes two kinds of errors. It must avoid errors of devoting processing resources to too many irrelevant changes (data overload) as well as errors of discarding too many potentially relevant changes as irrelevant (Woods, 1992). The critical cognitive function required to steer between these twin hazards for human, machine or joint cognitive systems, is attentional control (Gopher, 1991).

In addition, how does one handle data that comes in over time? How does one handle new events that occur while reasoning is underway, especially since these new events may reinforce current working hypotheses, elaborate on the current set of possible hypotheses, or completely overturn the current working hypothesis (Abbott, 1990)? What about the fact that there can be missing observations and erroneous signals?

The point of this step is to generate and update what are the significant findings to be explained.

STEP 2: H EXPLAINS D, THAT IS, H, IF TRUE, WOULD ACCOUNT FOR D

Note that there are multiple factors (automatic system responses, manual responses, and influences created by one or more faults) acting on the dynamic process. Multiple disturbances flow from a fault; multiple faults may be present; actions will be taken to mitigate the consequences of disturbances.

This means that multiple hypotheses (multiple factors) are needed to explain the set of findings D where each hypothesis accounts for a subset of the set of findings. Note that these hypotheses are about more than just the faults present in the system (e.g. responses to disturbances may be mis-executed). Thus, H is a set of hypotheses about what influences are acting (and have been acting) on the monitored process (although one can construe composite hypotheses, i.e. the hypothesis is that several hypothesized influences are acting on the system). The plausibility of hypotheses can change as new evidence comes in over time. Also, one may be able to identify different components of the set H at different stages of an evolving incident or at different confidence levels. Finally, one may need to characterize the components of the set H at a general level in order to act provisionally to compensate for the effects of the fault(s).

STEP 3: NO OTHER AVAILABLE HYPOTHESIS EXPLAINS D AS WELL AS H DOES

The absolutely critical word at this step in the formal characterization of abductive inference is *available* competitors (i.e. evaluate hypotheses relative to *available* competitors). This means that there is a process of hypothesis generation included in abductive inference in order to develop the set of competitors. Machine abduction systems generally finesse the hypothesis generation step through a mechanical process of hypothesis selection from a pre-enumerated set.

Because data come in over time in dynamic situations, it is possible that the initially most plausible hypotheses, given the findings available or most salient at that point, will turn out later in the evolving incident to be wrong or incomplete. We refer to these types of cases as garden path problems (Johnson, Moen *et al.*, 1988). This means that revision of hypotheses is a critical component of abductive inference in dynamic situations (De Keyser and Woods, 1990; Roth, Woods *et al.*, 1992).

CONCLUSION: THEREFORE, H IS PROBABLY TRUE

In dynamic situations, especially when there are high consequences to erroneous actions, the question of when to act arises. Should one act on the current best hypothesis or wait for more data, or wait for further diagnostic search to go on to investigate more possible alternative hypotheses? Remember in the types of domains that we are referring to, one can and often must act on a partial understanding of the situation. Thus, there can be provisional acceptance of an hypothesis set in order to act which may be different from commitment to a course of action.

Biases and errors in abduction in dynamic situations

The above characterization of abduction in the context of fault management of dynamic processes has implications for errors in abductive inference, for AI (Artificial Intelligence) performance systems targeted towards this class of cognitive situations, and for developing tools that support human abduction (cognitive aids).

While human reasoning biases and errors are a popular topic for deductive and inductive reasoning, very little has emerged about error/bias in abductive reasoning (Fraser, Smith *et al.*, 1992). The above characterization of abduction and dynamic processes shows how abductive inference can break down (i.e. lead to erroneous conclusions) at various points. The set of findings to be explained may be incomplete; incoherencies can arise in sorting out the pattern of multiple influences over time; hypothesis generation may be impoverished; revision of hypotheses may break down; imbalances can occur when juggling when to act versus waiting for more information (Woods, Roth *et al.*, 1987).

One example of error in abductive inference in dynamic fault management may be fixation errors or cognitive lockup (De Keyser and Woods, 1990) where the practitioner fails to revise an initial hypothesis despite the presence of cues (new or additional evidence) that should suggest that the earlier assessment is erroneous or that the state of the monitored process has changed, e.g. a new fault occurs (DeAnda and Gaba, 1991 report data on fixations in simulated anaesthesiology incidents). Note that AI abductive reasoners can fixate as well (Pople, 1985).

Monitoring the dynamic process to extract the findings to−be−explained can break down. The inherent variability of dynamic physical systems means that there are changes occurring all of the time in the monitored process that could be relevant, in principle. Out of all of the changing indications, fault managers must be able to recognize which represent anomalies. This is an example of a potential data overload situation where the critical cognitive activity is filtering the relevant indications from the irrelevant variations in the disturbed process (Woods and Roth, 1988; Doyle *et al.*, 1989; Woods, in press). A sensitive monitor in this kind of situation attempts to minimize two kinds of errors. It must avoid errors of devoting processing resources to too many irrelevant changes (data overload) as well as errors of discarding too many potentially relevant changes as irrelevant. The former type of error degrades performance because there are too many findings to be pursued and integrated into a coherent explanation of process state. Symptoms include loss of coherent situation assessment and response (e.g. vagabonding; Dorner, 1983), especially an inability to track the flow of events. The latter type of error degrades performance because the fault manager fails to pursue findings with high potential relevance and therefore sometimes fails to revise its/his/her situation assessment.

In human abduction, the problem is one of attentional control (Woods, 1992). Over control rejects too many potentially relevant changes as irrelevant. However, in trying to pursue and account for every change, attentional control becomes overwhelmed and breaks down. For machine reasoners, the problem becomes an inability to coherently track changes in the monitored process because evaluating every change invokes the full diagnostic reasoning capabilities of the machine. Building a coherent diagnostic assessment in the presence of the variability of dynamic physical systems becomes overwhelming. The solution has been to create implicitly or explicitly a monitor stage that identifies what are significant findings about the changing state of the monitored process that should trigger diagnostic cognitive work (Woods, Pople *et al.*, 1990; Abbott, 1990). Note that this solution imports the problem of attentional control into a machine reasoning context.

Errors also can intrude into abductive inference through failure to generate plausible alternative hypotheses. For human abduction, the reasoner can fail to 'call to mind' hypotheses; in other words, some hypotheses, which are known in principle, may not be activated and evaluated as plausible alternatives *in situ*. Calling to mind alternative hypotheses is a context-cued retrieval process in people. Note that what are available competitors may change as new incoming data serve as retrieval cues. In addition, the human may not know about all possibly relevant hypotheses. Machine abductive reasoners evaluate all of the hypotheses which it knows, i.e. are pre-enumerated, relative to a given finding; hypothesis generation failures occur outside in knowledge acquisition.

Another bias could occur in deciding whether to prefer single fault hypotheses over multi-fault accounts for the set of findings to be explained. A parsimony criterion is often interpreted as meaning that single fault hypotheses should always be preferred (Abbott, 1990; although see Peng and Reggia, 1990, for other definitions of parsimony). Adopting a theory of signal detection framework for considering the performance of diagnostic systems, Swets (1986) points out that one needs to distinguish between the sensitivity of the diagnostic system – its ability to correctly identify faults, and variations in a bias parameter that reflects tradeoffs relative to the costs and values placed on different types of potential errors and relative to the *a priori* probabilities associated with the situations to be identified independent of the level of sensitivity. If multi-fault cases are relatively likely (e.g. sensor error/failures may be relatively likely in combination with another fault when processes are heavily instrumented or prone to artifact) or if multi-fault cases are relatively undesirable (i.e. disasters in high consequence domains tend to be characterized by the combination of multiple factors rather than the occurrence of just one large failure, Reason, 1990), then the extreme bias setting – always prefer single fault hypotheses – is not justified normatively.

Aiding abduction in the context of fault management

With respect to cognitive aids to support human abduction in dynamic fault management, the analysis provided earlier suggests several approaches.

What are the findings to be explained (step 1)?

Representation aiding techniques could be used to help people identify and especially track changes in the set of findings to be explained (cf., Woods, in press, for a characterization of representation aiding).

One part of the cognitive work in AI diagnostic systems is to better identify, out of the large set of ongoing behaviours of the monitored process, which are anomalous, where an anomaly is some departure from, or contrast to, a reference condition such as goal state, limit violation or expected trajectory (Woods, Roth *et al.*, 1987). For example, qualitative reasoning may be used to develop a model of expected process behaviour and used to detect deviations from expected course, i.e. an important kind of anomaly that should trigger follow-up diagnostic cognitive work (step 2 in abductive inference) in order to develop potential explanations for the unexpected finding (Forbus, 1988; Woods, Pople *et al.*, 1990). However, a survey of human-intelligent system co-operation in fault management (Malin, Schreckenghost *et al.*, 1991; Woods, Potter *et al.*, 1991) found that the tools for co-operation generally failed to take advantage of the AI system's cognitive work on this score to enhance the human partner's view of the state of the monitored process. This suggests that more effort is needed to use intelligent data processing plus graphic techniques to develop representations of the monitored process that help the human practitioner recognize and track anomalies.

Another area where representation aiding may play a role relates to the disturbance propagation which is a characteristic result of faults in dynamic processes (Woods, 1988; Abbott, 1990). Temporal characteristics of disturbance propagation are highly informative about the nature of the fault, where to act to mitigate negative consequences, and where to act to break disturbance propagation. Human abduction may be aided through representations of the monitored process that reveal the temporally evolving pattern of disturbances – the human should be able to see disturbances grow, spread, and subside (Woods, Elm *et al.*, 1986; Potter and Woods, 1991).[2]

Another approach is to aid attentional control in a changing environment by improving the monitor's ability to discriminate between changes that are in fact irrelevant and should not invoke further cognitive processing versus changes that are of enough potential relevance to warrant the investment of additional processing resources. Field data suggest that effective alarm handling systems support this discrimination and that new types of alarm systems have failed when they undermine operators' ability to perform this discrimination (Woods, 1992, explores this approach in greater detail).

Multiple influences (step 2)

The state of the monitored process is the result of multiple influences that are or have been acting on it. An important part of fault management is to be able to separate and track the multiple factors at work (automatic system responses, manual responses and influences created by one or more faults), especially as they evolve dynamically. Does a new anomaly represent distur- bance propagation from the original fault or the effects of another break- down? Control influences inserted by one agent to mitigate the consequences of disturbances change the pattern of disturbances and symptoms, and will influence the generation and evaluation of fault hypotheses. For example, automatic actions to counteract a disturbance may inadvertently conceal the presence of a fault from supervisory control agents (Norman, 1990). Field studies of human performance at fault management show that a common strategy for follow up diagnostic search is to first check on recent control influences, interventions taken by some control agent. What control influ- ences are active? Did the influence I thought had initiated in fact occur, i.e. has an error in execution occurred? After updating and double checking this part of their situation assessment, diagnostic search switches its focus to possible process faults. Note that detecting/correcting errors in responding to the disturbances produced by faults is a part of effective fault management. This includes failures of automatic systems to respond as demanded, as well as human errors by omission or commission.

The above suggests a strategy for aiding dynamic abduction – helping people track the impact of these multiple influences. An integrated representation of

---

[2] The dynamic aspects of determining what are significant findings are the conceptual basis for two of the general principles of representation design: 1) effective representations should high- light contrasts so as to make it easy for observers to recognize the kind of anomaly present; 2) effective representations should highlight changes, behaviours and events, all of which refer to movement over time (Woods, in press).

known influences (e.g. automatic system response to a disturbance), hypothesized influences (possible faults), and the temporal change in the state of the monitored process can serve as the basic status board on which diagnostic reasoning takes place. The representation concepts noted in the previous section could serve as a basis for developing this integrated representation which would help to establish the common frame of reference that is essential for co-operative cognitive work (Woods and Roth, 1988; Hutchins, 1990).

Multiple hypotheses (step 3)

Since abduction involves evaluation of hypotheses relative to available competitors, the generation of possible alternative hypotheses is an area that can influence the quality of reasoning. Obviously, then, a strategy for aiding abduction is to aid hypothesis generation, i.e. to broaden the set of hypotheses under consideration as candidate explanations for the pattern of findings. There may be a variety of techniques which could be used in the representation of the monitored process, or the style of interaction between intelligent system and human operator, that could help people call to mind a broad set of plausible hypotheses.

Hypothesis generation may be an area where human–machine synergy may be much superior to the performance of either the human or the intelligent system alone. Unfortunately, in the dominant paradigm for coupling human and machine intelligence, the intelligent system generates its best estimate of *the* solution (Woods, 1986). There are reasons to suspect that this type of coupling does not broaden the set of hypotheses considered and may even narrow hypothesis generation from the point of view of the joint human–machine cognitive system.

Evaluation involves considering how well a called-to-mind hypothesis explains the pattern of findings. Thus, we could conceive of a representational window that helps the human supervisor monitor the intelligent system's evaluation by highlighting the relationship between a candidate hypothesis and the set of findings to-be-explained. This anomaly-hypothesis organization is a mapping of candidate hypothesis to findings that it explains out of the total set of findings to-be-explained (Potter and Woods, 1991). One can also represent this mapping in the other direction – organize around a given anomaly, the set of candidate hypotheses which could account for that finding. The anomaly-hypothesis organization may be one way to develop a common frame of reference to support co-operative human–machine fault management. This technique may also help provide 'on-line' rather than after-the-fact explanations of intelligent system conclusions, i.e. when the intelligent system presents conclusions, they already make sense to the human supervisor because of the previous context rather than they appear out-of-the blue as a surprise to be investigated.

The process of abductive reasoning is concerned with exploring different ways to put together subsets of findings which different subsets of hypothesized influences in order to generate a 'best' explanation. This suggests that one could extend the anomaly-hypothesis organization from a representation of diagnostic reasoning to an interactive diagnostic tool where the human and the intelligent system directly co-operate in exploring the implications of different ways to parse subsets of findings relative to possible influence patterns that would explain them. Some abductive machine reasoners are beginning to be capable of supporting such a style of interaction (Pople, 1985). This type of co-operation would seem to be too time consuming for many real-time process applications but, if the details of the interaction are designed adeptly in terms of human–computer interface, such a capability would afford the human diagnostician the ability to deal with many of the complexities that arise in dynamic situations, e.g. an initially plausible hypothesized influence pattern that should be revised as new evidence arrives.

Finally, there is the problem of how to help diagnosticians recognize that revision may be warranted given new incoming information, i.e. avoiding fixation errors (De Keyser and Woods, 1990).

When to act (step 4)?

It is important to remember that in fault management domains practitioners (commercial pilots, anaesthesiologists, mission control operators, nuclear power operators, etc.) are responsible, not just for device operation but also for the larger system and performance goals of the overall system. Given the high consequences that could occur, this responsibility casts a large and generally ignored shadow over cognitive processing.

These kinds of practitioners are responsible for action when the outcome is in doubt and consequences associated with poor outcomes are highly negative. In this charged cognitive environment, commitment to a course of action is a very important and under-appreciated cognitive task (Woods, O'Brien *et al.*, 1987). Bias towards taking corrective action with greater uncertainty or waiting for more data or diagnostic search before commitment is an important tradeoff at this stage of abductive reasoning, i.e. committing to a course of action (remember the theory of signal detection separation between a sensitivity parameter: how well the observer can discriminate different states and a bias parameter; how much to weight different possible outcomes given costs and benefits).

How do we aid this part of abduction in high consequence domains? To date we have evidence that intelligent systems designed as prosthetics create authority-responsibility double binds that undermine human performance on this part of abduction (cf., Woods, Johannesen, *et al.*, in press).

Aiding disturbance management

Faults present themselves as a cascade of disturbances in dynamic fault management applications. The consequences of this lead us to see the

under-appreciated cognitive task of disturbance management where diagnosis occurs as part of a larger context, where the expert practitioner must
maintain system integrity by coping with the consequences of faults (i.e. disturbances) through safing responses in parallel with untangling the causal
chain that underlies these disturbances in order to take corrective responses.
Abductive reasoning is a contributor to the disturbance management cognitive task. As a result, let us consider how to extend abductive reasoning away
from the exemplar of static troubleshooting of a broken device removed from
service and towards dynamic fault management where incidents extend,
develop and change over time. In this setting, abduction occurs as part of co-
operative ensemble. This leads us to consider abductive reasoning from a
human–computer co-operation or joint cognitive system point of view (Woods,
1986; Woods, Johannesen, *et al.*, in press). The joint cognitive system point
of view raises concerns about possibilities for error in abductive reasoning
and concerns about how to aid human abduction rather than the current
focus on machine performance systems.

It is hoped that this cognitive system model of dynamic fault management
will lead to the development of new techniques and concepts for aiding human
fault management and the design of intelligent systems in a variety of specific
application areas such as the air traffic system, automated flightdecks, space
control centres and other applications where dynamic fault management arises.
One implication is that the fault management support system should help the
operator see anomalies in the monitored process. Since anomalies are defined
as mismatches, the fault management support system should help the operator
see what specific mismatch or contrast is present. Since there are different
kinds of standards for process behaviour, e.g. target values, limit values,
automatic system response thresholds, intelligent system 'expectations' (in
the case of model-based AI systems), indications of an anomaly should include
the standard violated (Potter, Woods *et al.*, 1992).

Cognitive activities in fault management involve tracking the set of anomalies present in the process and their temporal inter-relationships. Fault management support systems and other representations of the behaviour of the
monitored process need to capture and highlight the aetiology of disturbance
evolution. The practitioner should be able to see the dynamics of anomalies
and the underlying disturbances in process functions, especially to see how
disturbances grow and subside in the face of safing/corrective responses
(Woods, Elm *et al.*, 1986; Potter and Woods, 1991). This information may be
very important in the diagnostic process and in the strategic allocation of
cognitive resources either to diagnostic search to identify the source of the
cascade of disturbances or to focus on coping/safing actions to protect important goals.

Again it is a fundamental feature of the disturbance management cognitive
task that diagnostic activities and information are intermingled with manual
and automatic responses to cope with the consequences of faults. How

the monitored process responds to these coping/safing actions provides information for the diagnostic process. Thus, it is important for a fault management support system to assist the practitioner to untangle the interaction between the influences of fault(s) and the influences of coping/safing actions taken by automatic systems or by some of the people involved.

Overall, note that the description of dynamic fault management outlines a series of cognitive demand factors – patterns of disturbances, temporal evolution of disturbances, interacting chains of disturbances derived from multiple faults, the interaction over time of abnormal and control influences, decompensation, the diagnostic role of interventions, etc. Each of these constitutes a kind of case or region within the problem space of fault management. Developing specific scenarios that sample from each of these regions in the problem space is a key first step for studies that will explore cognitive strategies in fault management or for evaluations to test the impact of particular alarm system concepts.

The evaluation of systems to aid fault management, either alarm systems or diagnostic systems cannot be carried out statically, but only against the backdrop of scenarios that sample the different dynamic aspects of disturbance management. For example, in one study of a computerized automated device (Moll van Charante, Cook *et al.*, 1993) it was found that alarms were remarkably common during device operation. In one sequence of about five minutes duration there were at least a dozen alarms from a single automated device. These alarms were not simply repeats of the same message but a variety of different messages associated with the sequence of events. It is important to note that – given the lack of feedback – when alarms sequences occurred it was very difficult for practitioners to determine what the automated device had been doing during the intervening period.

The cognitive systems model of dynamic fault management illustrates the intimate interaction between the cognitive demands of the task world, the cognitive attributes of the artifacts present, and the fact that a distributed set of agents do cognitive work in order to achieve goals. None can be understood except in relation to their interaction with the others as part of a cognitive system (Woods and Roth, 1988). Characterizing how they mutually influence each other is the cognitive systems research agenda; it can lead to the development of a relevant research base and can stimulate the development of effective cognitive tools (Woods, Johannesen, *et al.*, in press).

The demands of dynamic fault management go well beyond the simple strategy of developing automated fault diagnosis. Trying to finesse the complexities of disturbance management by allocating diagnosis to machine runs a great risk of neither successfully automating diagnosis nor developing effective means to support the human practitioner. The cognitive systems model sketched here hopefully indicates directions for the development of co-operative human–machine cognitive systems that will enhance performance at dynamic fault management.

Acknowledgements

Research support was provided by NASA under Grant NCA2–351 from the Ames Research Center, Dr Everett Palmer technical monitor, and under Grant NAG9–390 from the Johnson Space Center, Dr Jane Malin technical monitor. Many fruitful discussions with Jane Malin, Debbie Schreckenghost, Emilie Roth and Harry Pople have stimulated the ideas and relationships presented here. The work here would be impossible without the contributions of my colleagues at the Cognitive Systems Engineering Laboratory in studies of practitioners at work in disturbance management and in studies of systems that purport to aid fault management – Richard Cook, Scott Potter and Leila Johannesen.

References

Abbott, K.H., 1990, 'Robust fault diagnosis of physical systems in operation', Doctoral dissertation, State University of New Jersey, Rutgers.

Cook, R.I., Woods, D.D. and McDonald, J.S., 1991, Human performance in anesthesia: a corpus of cases, Cognitive Systems Engineering Laboratory Report, April.

Cook, R.I. and Woods, D.D., Operating at the 'sharp end': the complexity of human error, in Bogner, M.S. (Ed.) *Human Error in Medicine*, Hillsdale, NJ: Lawrence Erlbaum Associates, in press.

Davis, R., 1984, Diagnostic reasoning based on structure and behavior, *Artificial Intelligence*, **24**, 347–410.

DeAnda, A. and Gaba, D., 1991, Role of experience in the response to simulated critical incidents, *Anesthesia and Analgesia*, **72**, 308–15.

De Keyser, V. and Woods, D.D., 1990, Fixation errors: failures to revise situation assessment in dynamic and risky systems, in Colombo, A.G. and Saiz de Bustamante, A. (Eds) *Systems Reliability Assessment*, Dordrechts, The Netherlands: Kluwer Academic Publishers.

Dorner, D., 1983, Heuristics and cognition in complex systems, in Groner, R., Groner, M. and Bischof, W.F. (Eds) *Methods of Heuristics*, Hillsdale, NJ: Lawrence Erlbaum Associates.

Doyle, R.J., Sellers, S. and Atkinson, D., 1989, A focused, context sensitive approach to monitoring, in *Proceedings of the Eleventh International Joint Conference on Artificial Intelligence*, IJCAI.

Fraser, J., Smith, P.T. and Smith, J.N., 1992, A catalog of errors, *International Journal of Man–Machine Studies*, **37**, 265–307.

Forbus, K., 1988, Qualitative physics: past, present, and future, in *Exploring Artificial Intelligence*, San Mateo, CA: Morgan Kaufman.

Gaba, D., Maxwell, M. and DeAnda, A., 1987, Anesthetic mishaps: breaking the chain of accident evolution, *Anesthesiology*, **66**, 670–76.

Gopher, D., 1991, The skill of attention control: acquisition and execution of attention strategies, in *Attention and Performance XIV*, Hillsdale, NJ: Lawrence Erlbaum Associates.

Hutchins, E., 1990, The technology of team navigation, in Galegher, J., Kraut, R. and Egido, C. (Eds) *Intellectual Teamwork: Social and Technological Foundations of Cooperative Work*, Hillsdale. NJ: Lawrence Erlbaum Associates.

Johnson, P.E., Moen, J.B. and Thompson, W.B., 1988, Garden path errors in diagnostic reasoning, in Bolec, L. and Coombs, M.J. (Eds) *Expert System Applications*, New York: Springer-Verlag.

Josephson, J.R., 1991, Abduction: conceptual analysis of a fundamental pattern of inference, Laboratory for Artificial Intelligence Research Technical Report, The Ohio State University.

Klein, G., Orasanu, J. and Calderwood, R. (Eds) 1993, *Decision Making in Action: Models and Methods*, Norwood, NJ: Ablex.

Lind, M., 1991, Representations and abstractions for interface design using multi-level flow models, in Weir, G.R.S. and Alty, J.L. (Eds) *Human–Computer Interaction and Complex Systems*, London: Academic Press.

Malin, J., Schreckenghost, D., Woods, D., Potter, S., Johannesen, L. and Holloway, M., 1991, Making intelligent systems team players, NASA Johnson Space Center, TR-104738.

Moll van Charante, E., Cook, R.I., Woods, D.D., Yue, L. and Howie, M.B., 1993, Human–computer Interaction in context: physician interaction with automated intravenous controllers in the heart room, in Stassen, H.G. (Ed.) *Analysis, Design and Evaluation of Man–Machine Systems*, 1992, Pergamon Press.

National Transportation Safety Board, 1986, China Airlines B-747-SP, 300 NM north-west of San Francisco, *CA, 2/19/85, NTSB Report AAR-86/03*.

Norman, D.A., 1990, The 'problem' with automation: inappropriate feedback and interaction, not 'over-automation', *Philosophical Transactions of the Royal Society of London*, B 327: 585–93.

Peirce, C.S., 1955 (original 1903), Abduction and induction, in Buchler, J. (Ed.) *Philosophical Writings of Peirce*, Dover.

Peng, Y. and Reggia, J., 1990, *Abductive Inference Models for Diagnostic Problem Solving*, New York: Springer-Verlag.

Pople, Jr., H.E., 1985, Evolution of an expert system: from internist to caduceus, in De Lotto, I. and Stefanelli, M. (Eds) *Artificial Intelligence in Medicine*, New York: Elsevier Science Publishers B.V. (North-Holland).

Potter, S.S. and Woods, D.D., 1991, Event-driven timeline displays: beyond message lists in human-intelligent system interaction, in *Proceedings of IEEE International Conference on Systems, Man, and Cybernetics*, IEEE.

Potter, S.S., Woods, D.D., Hill, T., Boyer, R. and Morris, W., 1992, Visualization of dynamic processes: Function-based displays for human-intelligent system interaction, in *Proceedings of IEEE International Conference on Systems, Man, and Cybernetics*, IEEE.

Rasmussen, J., 1986, *Information Processing and Human-Machine Interaction: An Approach to Cognitive Engineering*, Amsterdam: North-Holland.

Reason, J., 1990, *Human Error,* Cambridge: Cambridge University Press.

Reiersen, C.S., Marshall, E. and Baker, S.M., 1988, An experimental evaluation of an advanced alarm system for nuclear power plants, in Patrick, J. and Duncan, K. (Eds) *Training, Human Decision Making and Control*, New York: North-Holland.

Roth, E.M., Woods, D.D. and Pople, Jr., H.E., 1992, Cognitive simulation as a tool for cognitive task analysis, *Ergonomics*, **35**: 1163–98.

Sarter, N. and Woods, D.D., Pilot interaction with cockpit automation II: an experimental study of pilot's model and awareness of the flight management system, *International Journal of Aviation Psychology*, in press.

Schwid, H.A. and O'Donnell, D., 1992, Anesthesiologist's management of simulated critical incidents, *Anesthesiology*, **76**: 495–501.

Swets, J.A., 1986, Form of empirical ROCs in discrimination and diagnostic tasks: implications for theory and measurement of performance, *Psychological Bulletin*, **99**; 181–98.

Woods, D.D., 1986, Paradigms for decision support, in Hollnagel, E., Mancini, G. and Woods, D. (Eds) *Intelligent Decision Support in Process Environments*, New York: Springer-Verlag.

Woods, D.D., 1988, Coping with complexity: the psychology of human behavior in complex systems, in Goodstein, L.P., Andersen, H.B. and Olsen, S.E. (Eds) *Tasks Errors, and Mental Models*, London: Taylor & Francis.

Woods, D.D., 1992, The alarm problem and directed attention, *Cognitive Systems Engineering Laboratory Report*, CSEL 92-TR-05, October.

Woods, D.D., Towards a theoretical base for representation design in the computer medium: ecological perception and aiding human cognition, in Flach, J., Hancock, P., Caird, J. and Vicente, K. (Eds) *An Ecological Approach to Human–Machine Systems I: A Global Perspective*, Erlbaum, in press.

Woods, D.D. and Hollnagel, E., 1987, Mapping cognitive demands in complex problem solving worlds, *International Journal of Man–Machine Studies*, **26**, 257–75 (Also in Gaines, B. and Boose, J. (Eds) 1988, *Knowledge Acquisition for Knowledge Based Systems*, London: Academic Press.

Woods, D.D., Johannesen, L., Cook, R.I. and Santen, N., *Behind Human Error: Cognitive Systems, Computers Hindsight*, Crew Systems Ergonomic Information and Analysis Center, Dayton OH (State of the Art Report), in press.

Woods, D.D. and Roth, E.M., 1988, Cognitive systems engineering, in Helander, M. (Ed.) *Handbook of Human–Computer Interaction*, New York: North-Holland.

Woods, D.D., Elm, W.C. and Easter, J.R., 1986, The disturbance board concept for intelligent support of fault management tasks, in *Proceedings of the International Topical Meeting on Advances in Human Factors in Nuclear Power*, American Nuclear Society/European Nuclear Society.

Woods, D.D., O'Brien, J. and Hanes, L.F., 1987, Human factors challenges in process control: the case of nuclear power plants, in Salvendy, G. (Ed.) *Handbook of Human Factors/Ergonomics*, New York: Wiley.

Woods, D.D., Pople, H.E. and Roth, E.M., 1990, Cognitive environment simulation as a tool for modeling human performance and reliability, Vol. 2, US Nuclear Regulatory Commission, NUREG-CR-5213.

Woods, D.D., Potter, S.S., Johannesen, L. and Holloway, M., 1991, Human interaction with intelligent systems: trends, problems, new directions, *Cognitive Systems Engineering Laboratory Report*, CSEL 91-TR-01, February.

Woods, D.D., Roth, E.M. and Pople, Jr., H.E., 1987, Cognitive environment simulation: an artificial intelligence system for human performance assessment Vol. 2, US Nuclear Regulatory Commission, NUREG-CR-4862.

6

Alarm initiated activities

Neville Stanton

Introduction

The need to examine alarm handling behaviour stems from difficulties experienced by operators with industrial alarm systems (Pal and Purkayastha, 1985). Across a range of industrial domains, alarm systems appear to place the emphasis on detection of a single event, rather than on considering the implications of the alarm within the task (Stanton, 1993). Therefore, current industrial systems do not appear to make optimum use of human capabilities which could improve the overall human supervisory control performance (Sorkin, 1989). This is desirable because we are unlikely to remove human operators from the system. This would require a level of sophistication not possible in the foreseeable future. However, the reluctance to leave a machine in sole charge of 'critical' tasks is likely to mean that human operators will still be employed in a supervisory capacity because of concern about breakdown, poor maintenance, as well as ethical concerns. Therefore we need to capitalize on the qualities that operators bring to the 'co-operative endeavour' of human–machine communication. Alarm problems are further confused by the inadequacies of peoples' understanding of what constitutes an 'alarm' (Stanton and Booth, 1990). Most definitions concentrate on a subset of the qualities or properties, for example 'an alarm is a significant attractor of attention' or 'an alarm is a piece of information'. In fact, an alarm may be considered from various perspectives (Singleton, 1989), which need to be integrated into one comprehensive definition if the term is to be understood in its entirety. An 'alarm' should be defined within a systems model and consider how each of the different perspectives contribute to the interpretation of the whole system (Stanton, Booth *et al.*, 1992). In this way, one may examine the role of the human operator in response to alarm information, in order to develop a model of alarm handling that will ultimately influence alarm system design. A model may be considered to be a description or representation of a process that enables analysis of its form to be undertaken. A model of alarm handling is necessary to guide research, so that we may ask appropriate questions and utilize suitable empirical techniques to yield answers.

The development of models to understand human behaviour within complex systems is not a new endeavour (Edwards and Lees, 1974; Broadbent, 1990). It has been the domain of cognitive psychologists and human factors researchers alike. Models serve practical purposes, such as:

- a framework to organize empirical data;
- a prompt for investigation;
- to aid design solutions;
- to compare with actual behaviour;
- to test hypotheses and extrapolate from observable inferences;
- to measure performance;
- to force consideration of obscure or neglected topics.
  (Pew and Baron, 1982).

Models may be coarsely split into two types: quantitative and qualitative. Quantitative models are computational, (for example: simulations and analytic or process models) whereas qualitative models are descriptive. Quantitative models can produce mathematically precise estimates of performance (Broadbent, 1990; Elkind, Card *et al.*, 1990), but they are limited to use in highly specialized and restricted domains. Often the lack of hard data to put into a quantitative model of human behaviour means that one must first develop qualitative models. These serve as a basis for collecting the necessary empirical data that could eventually provide the information for a quantitative model.

Many qualitative models of human intervention in control room incidents have been proposed (Edwards and Lees, 1974; Rasmussen, 1976; Rouse, 1983; Hale and Glendon, 1987; Swain and Weston, 1988). The best known of these are the models of Rouse (1983) and Rasmussen (1976, 1983, 1984, 1986). Rasmussen's Skill–Rule–Knowledge (SRK) framework is extensively cited in the literature, and has been accepted as 'the industry standard' (Reason, 1990). The SRK framework distinguishes between three levels of performance that correspond with task familiarity. At the lowest level, skill-based performance is governed by stored patterns of proceduralized instructions. At the next level, behaviour is governed by stored rules, and at the highest level, behaviour is governed by conscious analytical processes and stored knowledge. Pew, Miller *et al.* (1982) comment on the strengths of Rasmussen's framework which they present as a decision making model which contains three essential elements that are consistent with human problem solving: data processing activities, resulting states of knowledge and shortcuts in the 'stepladder' model (discussed next).

Reason (1990) commented on Rasmussen's eight stages of decision making for problem solving: activation, observation, identification, interpretation, evaluation, goal selection, procedure selection and activation. He suggested that Rasmussen's major contribution was to have charted the shortcuts that human decision makers take in real situations (i.e. the stepladder model) which result in 'highly efficient, but situation-specific stereotypical reactions'. Pew and Baron (1982) provides an example of problem detection, for which the operator collects limited data and may immediately conclude that a specific control action must be executed (skill-based behaviour). Alternatively, the operator may additionally identify the system state and then select and execute a procedure that results in an action sequence (rule-based behaviour). Finally when the circumstances are new or the specific combination of circumstances does not match known ones, then the whole range of problem solving behaviour is called forth (knowledge-based behaviour). Reason (1988b) suggests that most incidents are likely to require this last type of behaviour, because although they may start in a familiar way they rarely develop along predictable lines. It is this unpredictable development that gives the greatest cause for concern, particularly when the true nature of the incident departs from the operator's understanding of it (Woods, 1988). As Reason (1988b) notes:

> each incident is a truly novel event in which past experience counts for little, and where the plant is returned to a safe state by a mixture of good luck and laborious, resource limited, knowledge-based processing.

From an extensive review of the literature on failure detection, fault diagnosis and correction, Rouse (1983) identified three general levels of human problem solving, namely:

- recognition and classification;
- planning; and
- evaluation and monitoring.

Within each of these levels Rouse assigns a three stage decision element to indicate whether the output of each stage is skill-based, rule-based or knowledge-based, rather like Rasmussen's framework. Firstly it is assumed that the individual is able to identify the context of the problem (recognition and classification), and then is able to match this to an available 'frame'. If a 'frame' does not exist then the individual has to resort to first principles. At the planning level, the individual must decide if a known procedure can be used, or whether alternatives have to be generated. Problem solving is generated at the lowest level where plans are executed and monitored for success. Familiar situations allow 'symptomatic' rules (i.e. rules based upon identifying familiar plant symptoms), whereas unfamiliar situations may require 'topographic' rules (i.e. rules based upon an understanding of the physical topography of the plant and the cause–effect relationships of the components). However, it has been argued that human problem solving is characterized by its opportunistic nature, rather than following a hierarchical information flow (Rouse, 1983; Hoc, 1988), with all levels being employed simultaneously. This would suggest a problem-solving heterarchy utilizing parallel processing. Therefore, the SRK model is not without its critics. Bainbridge (1984) suggests that at best it presents an oversimplified account of cognitive activity, and that at worst the inferences drawn may be wrong. Her main criticisms may be summarized as:

- a confusion of the terminology;
- a failure to represent all aspects of human behaviour;
- missing important aspects for the understanding of human cognition.

She warns of the danger of a strict application of the SRK framework which might restrict the flexibility of human behaviour, for example, by providing displays that can only be used for limited purposes. However, she does accept that it provides the basic idea of cognitive processes. Most of the criticism of the SRK framework has arisen either from a misunderstanding of the original intention, which was to provide a framework rather than a grand psychological theory, or from inappropriate application (Goodstein, Andersen *et al.*, 1988). Thus within its accepted limitations, it has remained robust enough to be considered a working approximation to human cognitive activities and allows for some prediction and classification of data.

Much of the attention paid to the SRK framework has been in the domain of human supervisory control, and Reason (1988b) presented the 'catch-22' of such systems.

- The operator is often ill-prepared to cope with emergencies, because the relatively low frequency of the event means that it is likely to be outside his/her experience. Moreover, high levels of stress are likely to accompany the emergency, making the operator's task more difficult.
- It is in the nature of complex, tightly-coupled, highly interactive and partially understood process systems to spring nasty surprises (Perrow, 1984).

The first point was made eloquently by Bainbridge (1983) in her discussion of the 'ironies of automation'. In the design of complex systems, engineers leave the tasks they cannot automate (or dare not automate) to the human, who is meant to monitor the automatic systems, and to step in and cope when the automatic systems fail or cannot cope. However, an increasing body of human factors knowledge and research suggests that the human is poor at monitoring tasks (Moray, 1980; Wickens, 1984; Moray and Rotenberg, 1989). When the humans are called to intervene they are unlikely to do it well. In other words, removing the humans from control is likely to make the task harder when they are brought back in (Hockey, Briner *et al.*, 1989). It has been suggested that diagnosis and control behaviour and quite different (Wickens, 1984). However, diagnosis behaviour is likely to be (at least in part) adapted to the way in which the information is presented to the operator and vice versa. Therefore emphasis needs to be put on understanding how

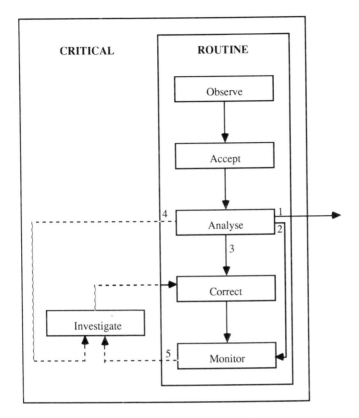

Figure 6.1   Model of alarm initiated activities.

the operator uses and processes the information, and to relate this understanding back to human cognitive activity in fault management in general.

Model of alarm initiated activities

The following model was constructed by Stanton (1992). As shown in Figure 6.1, it highlights the difference between routine incidents involving alarms (plain lines) and critical incidents involving alarms (dotted lines). The distinction between 'routine' and 'critical' is determined by the operator in the course of alarm handling. Although there are common activities to both types of incident (Figure 6.1), critical incidents require more detailed investigations. It is proposed that the notion of alarm initiated activities (AIA) is used to describe the collective of these stages of alarm event handling. The term 'activities' is used here to refer to the ensuing cognitive modes as well as their corresponding behaviours, both of which are triggered by alarms. The AIA are assumed to be distinctly separate activities to 'normal' operation in supervisory control tasks.

Table 6.1   Example of alarm initiated activities

Event	Outcome	AIA
1 Pump temperature exceeds alarm threshold.	'Pump ABC Temp High' alarm flashes accompanied by tone	
2 Operator hears alarm tone	Operator looks at alarm panel	Observe
3 Operator presses 'alarm accept' key	Alarm stops flashing and tone is silenced	Accept
4 Operator reads alarm message	(a) ignore alarm (b) monitor situation (c) reset alarm (d) investigate	Analyse
5 Operator investigates cause of pump ABC overheating	Operator finds valve XYZ closed	Investigate
6 (a) Operator opens valve XYZ (b) Operator stops pump ABC	(a) Valve XYZ opens (b) Pump ABC stops	Correct
7 Operator intermittently checks pump ABC	Pump ABC temperature eventually comes below threshold for 'Temp High' alarm	Monitor
8 Operator resets 'Pump ABC Temp High' alarm	'Pump ABC Temp High' alarm returns to non-active state	Correct

Typically control desk engineers (CDEs) report that they will observe the onset of an alarm, accept it and make a fairly rapid analysis of whether it should be ignored (route 1), monitored (route 2), dealt with superficially (route 3) or require further investigation (route 4). Then, even if they feel that it may require further investigation, they may still try to correct and cancel it (route 3) just to see what happens. If it cannot be cleared, then they will go into an ivestigative mode to seek the cause (route 5). Then in the final stage the CDEs will monitor the status of the plant brought about by their corrective actions. The need to resort to the high cognitive level 'investigation' is what distinguishes critical from routine incidents. The stages of activity may be considered with the help of an example of alarm handling taken from a manufacturing industry (Table 6.1).

Consider the filling of a tank from a storage vessel through a pipe with a valve and pump in-line. The operator in the control room is busy with various aspects of the task, such as the setting up of equipment further on in the process when he/she hears an audible alarm (event 2 in Table 6.1). The alarm is acknowledged by the cancellation. The operator now has a variety of options, as it is not yet known why the alarm telling the operator that the pump has overheated was triggered. There are a number of plausible explanations, such as:

1. there is a physical fault with the pump;
2. the storage vessel is empty;

3. the supply pipe is blocked or leaking; or
4. the valve is closed.

Given these degrees of uncertainty, there are several different remedial actions open to the operator as shown by outcomes to event 4. One path to saving the pump might be to stop it running (event 6b). Alternatively the operator may attempt to find the cause of overheating, which may be due to the valve not being opened before the pump was switched on. This may lead the operator to open the valve (event 6a) and then intermittently check the status of 'pump ABC' (event 7). Eventually the alarm will change status and enable the operator to reset it (event 8).

The above is an idealized description of a successful path through the series of events, and as such gives a simplified account of the true nature of the task. It assumes that the operator was successfully able to identify the reason for the alarm, although the alarm cue did not directly point to it. In this case there was a variety of plausible alternatives, each of which would require investigation. Whether or not exhaustive discounting actually takes place depends on the operator being able to bring them to mind.

The criteria for defining success are also ambiguous. If the operator stops the pump (event 6b), this would lead to the alarm being cleared, thus providing the opportunity to route the product through another pipe to fill the tank. Such a strategy would, perhaps, have been equally successful as the first alternative selected. In reality there may be many different possible courses of action competing for the operator's time and attention depending on the number of active alarms. The task is made even more difficult by the fact that alarms may also be grouped by events, and be interdependent on each other. This is particularly true in closely coupled systems (Perrow, 1984) with feedback loops. Such grouping can make the task of distinguishing cause and effect very difficult and, in turn, add to the inherent ambiguities described earlier.

As the example demonstrates, an alarm handling sequence can be described as consisting of a number of generic activity stages. The activities are illustrated in the AIA (alarm initiated activities) column of Table 6.1. Studying the alarm handling activities employed by operators might give some indication of how best to design alarm systems. This argument will be developed within the chapter.

Therefore, a consideration of the literature is required to make further inference about the requirements of these stages of handling. These AIAs will provide the framework of the review and guide subsequent research. The review is presented in the following sections: observe, accept, analyse, investigate, correct and monitor.

Observe

The observe mode is characterized by the initial detection of abnormal plant conditions. Detection is the act of discovering any kind of undesired deviation(s) from normal system operations (Johannsen, 1988). Bainbridge (1984) suggests that there are three main ways of detecting abnormal plant conditions:

- responding to an alarm;
- thinking of something that needs to be checked;
- incidentally noticing that something is wrong whilst attending to something else.

Failure to detect an abnormal situation may occur for a number of reasons (Moray, 1980):

- the relevant variable is not displayed;
- the signal to noise ratio is too low;
- the expectation of the operators leads to a misinterpretation of the information;
- the information may be ignored due to attention being directed on other variables;
- there may be too much information.

Under normal conditions Moray suggests that most systems are adequate to allow visual scanning to support monitoring tasks. However, when very rapid changes occur the task becomes very difficult. Prolonged activity of this kind is likely to reduce the efficiency of human cognitive activities as

> several concurrent activities may compete for access to a particular (cognitive) 'resource' . . . the cost of errors may be very great.
>
> Hockey, Briner *et al.* (1989)

Counter to an intuitive notion of the control task, Moray (1980) suggests that the better the system is known to an operator, the less likely he/she will discover an abnormal state. He implies that this is due to the reliance of the operator on past experience and the correlation between variables to predict future states. This leads to a failure to observe current values. Therefore abnormal values are undetected. This proposition is similar to the observations of Crossman and Cooke (1974) who noticed that skilled tracking behaviour was primarily 'open-loop'. Tracking is compensatory (that is it occurs after the event), therefore when dealing with highly familiar data the human is likely to fill in the gaps or miss the data. Reason (1990) suggests that as fault detection moves from being knowledge-based to becoming skill-based, it is likely to suffer from different types of error. Reason proposes that skill-based behaviour is susceptible to slips and lapses whereas knowledge-based behaviour is susceptible to mistakes.

In a series of experiments aimed at investigating fault detection in manual and automatic control systems, Wickens and Kessel (1981) concluded that automating the system does not necessarily reduce the mental workload of the human controller. Firstly they noticed a paradox of task operation. In manual control, operators are able to continually update their 'model' of the system, but are also required to perform two tasks: control and detection. Whereas in automatic control they had only the detection task, but were not 'in-loop' to update their 'model'. This means that removing the human from the control loop may reduce the attention paid to the system state. Wickens and Kessel suggest that whether the manual or automatic control task performance was superior would depend largely upon the relative workload, i.e. under some conditions workload might favour manual control and in others workload might favour automatic control. Automation shifts the locus of the information processing demands. In manual control, the emphasis is primarily on 'responding', whereas in automatic control the demands are primarily located in 'perception' and 'central processing'. Under the SRK framework the shift is from skill-based behaviour to knowledge- and rule-based behaviour.

Wickens and Kessel also suggest a 'fragility' of failure detection performance as:

- it cannot benefit from borrowed resources of responding;
- it deteriorates when responding demand is increased.

In summary, it appears that detection has the 'worst of both worlds'. This may represent an intrinsic characteristic of detection tasks in general.

In a series of investigations into fault management in process control environments, Moray and Rotenberg (1989) observed that subjects:

- display cognitive lockup when dealing with a fault;
- prefer serial fault management;
- experience a time delay between noticing a fault and dealing with it.

Moray and Rotenberg noticed that when dealing with one fault their subjects would not take action on another. This is linked to the preference for dealing with faults serially, rather than concurrently. Moray and Rotenberg were however, unable to distinguish between cause and effect, i.e. whether cognitive lockup leads to subjects dealing with faults serially or vice versa. In process systems, serial fault management may not produce optimum process performance, but it may make task success more likely, as interruptions in fault management (to deal with other faults) may cause the human operator to forget important aspects of the first task that was being worked on. The data collected by Moray and Rotenberg can explain the time delay between looking at a fault and dealing with it. The data showed that a fault is examined many times before intervention is initiated. Their eye-movement data demonstrate that just because operators are not actively manipulating controls we cannot assume that their task load is low. Moray and Rotenberg's data suggest that the operator is actively processing information even in apparently non-active periods. They claim that an operator might observe an abnormal value, but fail to take action for at least three reasons:

- the evidence was not strong enough to lead to a diagnosis for appropriate action;

- the operator was already busy dealing with another fault and wishes to finish that problem before starting a new one;
- although the abnormal value was observed, it was not perceived as abnormal.

They conclude from their data that the second of these proposals appears most likely in their investigation. The locking-up of attention is a phenomenon that has been repeatedly reported in the literature (e.g. Moray and Rotenberg, 1989; Hockey, Briner *et al.*, 1989; Wickens, 1984) and appears to be a intrinsic characteristic of human cognitive processing. As Wickens (1984) expresses it:

> . . . it is reasonable to approximate the human operator as a single-channel processor, who is capable of dealing with only one source of information at a time.

The irony of attracting the operator's attention to the new alarm information is that successful attraction will necessarily mean distracting the operator from other aspects of the task. The interruption may not be welcome as it may interfere with some important operation. Therefore the alarm system needs to show that a problem is waiting to be dealt with, rather than forcing the operator to deal with it unless the alarm merits immediate action, and enable the operator to distinguish between alarms that relate to separate events. Moray and Rotenberg (1989) report that the probability of looking at a fault and dealing with it may be described in terms of a logarithmic relationship between probability of detection and time since its occurrence.

Accept

The acceptance of an alarm is taken to be acknowledgement or receipt. This is normally a physical action that takes the alarm from its active state to a standing state. Jenkinson (1985) proposed that audible and visual cues should be combined to reduce the visual search task, as the operator has to move within the workspace, and visual information alone is insufficient. Normally the receipt of an alarm is accompanied by the silencing of the audible cue, and a change in some aspect of the visual coding, such as from flashing to illuminated. However, this change in visual and auditory state may make it difficult to tell when an alarm has been accepted. For example, in an annunciator or mimic display, once the flashing code has stopped there may be no means of recording the time or order of occurrence of the alarm. So by accepting it, the operator loses some information about the alarm that may be essential for the subsequent AIAs, (such as 'analyse' or 'investigate') to be performed effectively. However, the alarm may be considered to be in one of four possible states:

- not activated;
- activated but not accepted;

᾿ accepted but not reset;
• reset.

Resetting an alarm is the acknowledgement by the operator that the initiating condition is no longer present. It extinguishes the alarm, returning it to its first state: not activated. The indication that an alarm is waiting to be reset is normally in the form of a marker or code (Jenkinson, 1985) to inform the operator of its new state.

The designers of alarm systems have to consider whether to allow group acknowledgement of alarms, or to insist on each alarm being acknowledged individually. Unfortunately the literature is inconclusive. Group acknowledgement of alarms may cause the operators to deal inadvertently with a signal (Kragt and Bonten, 1983) but single acknowledgement may fare no better (Kortlandt and Kragt, 1980). With group acknowledgement it is possible that the operator could miss a signal by accepting *en masse* and scan the alarm list or matrix. However, in periods of high alarm activity it is likely that single acknowledgement actions will resemble group acknowledgement, as the operator repeatedly presses the 'accept' key without reading the alarm message (Stanton, 1992). Reed and Kirwan (1991), however, describe the development of an alarm system that requires operators to accept each alarm individually.

Under certain operational situations up to 200 alarms could be presented. They claim that the simplicity of the task will mean that single acknowledgement of each of the 200 alarms will not be unduly problematic. What they do not acknowledge is that tying the operators up in this simple acceptance task prevents them from moving further on in the alarm initiated activities. This could become a problem if there are other critical failures within the process that are hidden within the 200 alarms presented. Further, an operator may sometimes accept a signal just to get rid of the audible signal (Kragt and Bonten, 1983; Sorkin, 1989). This presents a paradox in design, because the operator is made aware of a change in the process state by the presence of the signal attracting attention. Failure to attend to the alarm will mean that it is impossible to pass this information on to the subsequent stages of AIAs. Masking of a fault may result from too many alarms. This was the most often cited reason for missing alarms in recent studies (Stanton, 1993).

Analyse

Analysis may be considered to be the assessment of the alarm within the context of the task that is to be performed and the dynamics of the system. Analysis appears to involve a choice of four options (ignore alarm, monitor situation, deal with alarm superficially or investigate cause) and therefore involves some rudimentary search of context to reach an appropriate judgement. Easterby (1984) proposed that a variety of psychological processes are used by an operator in control of a machine, such as: detection, discrimination, identification, classification, recognition, scaling, ordering and sequencing. He suggested that the control panel may be considered as a map of the operator's task:

> the display must therefore define the relationships that exist between the machine elements, and give some clues as to what to do next.

This is essentially the operator's task in analysis: to decide what to do next. Operators are often required to search for the relevant information to base their decisions on, as in VDU-based control systems the information is not necessarily available immediately, and can only be obtained after request (Kragt and Bonten, 1983).

From the reported behaviours of plant operators, the results of the analysis stage of AIAs determine the future course of action: ignoring the alarm, monitoring the system, making superficial corrective actions to cancel the alarm, or going into an investigative mode. This puts an emphasis on the alarm to convey enough information to make this decision without involving the operators in too much effort as there may be other demands upon their attention. To some extent operators may be aided in the task by a current awareness of the plant state. For example, if they know that a part of the plant is in maintenance, then they are unlikely to be surprised that the value of a particular variable is outside its normal threshold. Alternatively if they are tracking the development of an incident, an alarm may confirm their expectations and therefore aid diagnosis. However, it is also possible that the operators may wrongly infer the true nature of the alarm leading to an inappropriate analysis and subsequent activity. It is important to note that the presence of the alarm by itself may not directly suggest what course of action is required, but only reports that a particular threshold has been crossed. In the search for the meaning of the alarm, the manner in which it is displayed may aid or hinder the operator. For example alarm lists show the order in which the alarm occurred; alarms within mimic displays map onto the spatial representation of the plant, and annunciator alarms provide the possibility for pattern recognition.

These different ways of presenting alarm information may aid certain aspects of the operator's task in analysis, such as indicating where the variable causing the alarm is in the plant; what the implications of the alarm are; how urgent the alarm is, and what should be done next. Obviously different types of information are conveyed by the different ways to present alarm information mentioned (lists, mimics and annunciators). The early classification process may be enhanced through pairing the visual information with auditory information such as tones or speech. Tones are abstract and would therefore require learning, but may aid a simple classification task such as urgency (Edworthy and Loxley, 1990).

Tones provide constant information and are therefore not reliant on memory for remembering the content of the message. They are reliant on memory for recalling the meaning of the message. Whereas speech is less abstract and rich in information, it is varied and transitory in nature, so whilst it does have the possibility of providing complex information to the operator in a 'hands-free eyes-free' manner, it is unlikely to find favour as an alarm medium in process control (Baber, 1991).

It has been speculated that text and pictures are processed in a different manner (Wickens, 1984), and there are alternative hypotheses about the underlying cognitive architectures (Farah, 1989). Wickens' dual face multiple resource theory and stimulus–cognitive processing–response (SCR) compatibility theory offer an inviting, if mutually irrefutable, explanation of information processing. Wickens' theories predict that the modality of the alarm should be compatible with the response required provided that the attentional resources for that code are not exhausted. If attentional resources for that code are exhausted, then another input modality that does not draw on the same attentional resources should be used. Despite the attraction of Wickens' explanation, based on a wealth of data involving dual task studies, there is still some contention regarding the concept of separate information processing codes. Farah (1989) draws a clear distinction between the three main contending theoretical approaches to the representation of peripheral encoding and internal cognitive processing. First, Farah suggests that although encoding is specific to the input modality, internal processing shares a common code. Second, the single code approach is favoured by the artificial intelligence community, probably because of the computational difficulties of other approaches (Molitor, Ballstaedt *et al.*, 1989). Alternatively (third) the 'multiple resource' approach proposes separate encoding and internal processing codes (Wickens, 1984). Farah (1989) suggests that recent research points to a compromise between these two extremes.

Recent studies have shown that a combination of alphanumeric and graphic information leads to better performance than either presented alone (Coury and Pietras, 1989; Baber, Stammers *et al.*, 1990) It might similarly be speculated that the combination of codes in the correct manner may serve to support the analysis task. The model of AIAs implies that different aspects of the code might be needed at different points in the alarm handling activity. Thus the redundancy of information allows what is needed to be selected from the display at the appropriate point in the interaction. The type of information that is appropriate at any point in the interaction requires further research.

Investigate

The investigative stage of the model of AIAs is characterized by behaviour consistent with seeking to discover the underlying cause of the alarm(s) with the intention of dealing with the fault. There is a plethora of literature on fault diagnosis, which is probably in part due to the classical psychological research available on problem solving. The Gestalt psychology views provide an interesting but limited insight into problem solving behaviour, confounded by vague use of the terminology. Research in the 1960s was aimed at developing an information processing approach to psychology in general, and to problem solving in particular, to:

> ... make explicit detailed mental operations and sequences of operations by which the subject solved problems.
>
> Eysenck (1984)

A closer look at research from the domain of problem solving illustrates this clearly. Problem solving may be considered analogous to going through a maze, from the initial state towards the goal state. Each junction has alternative paths, of which one is selected. Moving along a new path changes the present state. Selection of a path is equivalent to the application of a number of possible state transforming operations (called operators). Operators define the 'legal' moves in a problem solving exercise, and restrict 'illegal' moves or actions under specific conditions. Therefore a problem may be defined by many states and operators, and problem solving consists of moving efficiently from our initial state to the goal state by selecting the appropriate operators. When people change state they also change their knowledge of the problem. Newell and Simon (1972) proposed that problem solving behaviour can be viewed as the production of knowledge states by the application of mental operators, moving from an initial state to a goal state. They suggested that problem solvers probably hold knowledge states in working memory, and operators in long term memory. They problem solver then attempts to reduce the difference between the initial state and the goal state by selecting intermediary states (subgoals) and selecting appropriate operators to achieve these. Newell and Simon suggest that people move between the subgoal states by:

• noting the difference between present state and goal state;
• creating a subgoal to reduce the difference; and
• selecting an operator to achieve this subgoal.

Thus it would appear that the cognitive demand of the task is substantially reduced by breaking the problem down, moving towards the goal in a series of small steps. A variety of computer-based systems have been produced in an attempt to model human problem solving, but none have provided a wholly satisfactory understanding. This is not least because they are unable to represent problem solving in everyday life, and computer models rely on plans, whereas actions may be performed in a number of ways. As Hoc (1988) proposes:

> A problem will be defined as the representation of a task constructed by a cognitive system where this system does not have an executable procedure for goal attainment immediately at its disposal. The construction of a task, representation is termed understanding, and the construction of the procedure, problem solving.

This means that the same task could be termed a problem for some people, but not for others who have learned or developed suitable procedures (Moran,

1981). The difficulty in analysing problem solving is the human ability to perform cognitive activity at different levels of control at the same time. Rasmussen's SRK framework is useful in approximating these levels, but the entire activity leading to a goal can seldom be assigned to one, and usually occurs at all levels simultaneously. Hoc (1988) sees problem solving as involving two interrelated components: problem understanding (the construction of a coherent representation of the tasks to be done) and procedure searching (the implementation of a strategy to find or construct a procedure). This suggests that there is an 'executive controller' of the problem solving activities which directs the choices that are taken (Rouse, 1983). Planning is the guiding activity that defines the abstract spaces and is typically encountered in problem solving. Hoc (1988) believes that planning combines top–down components (creating new plans out of old ones) with bottom–up components (elaborating new plans or adapting old plans). Thus he suggests that an information representation that supports the shift between these components would result in more efficient strategies. Human factors is essentially about the design of environments that suit a wide range of individuals. Therefore presentation of information that only suits one strategy, or particular circumstances, is likely to frustrate the inherent variation and flexibility in human action.

Landeweerd (1979) contrasts diagnosis behaviour with control, proposing that, in control, the focus of attention is upon the forward flow of events, whereas diagnosis calls for a retrospective analysis of what caused what. Wickens (1984) widens the contrast by suggesting that the two tasks may be in competition with each other for attentional resources and that the two phases of activity may be truly independent. However, whilst diagnosis certainly does have a retrospective element in defining the problem, it certainly has a forward looking element of goal directed behaviour in correcting the fault. Landeweerd (1979) suggests that the type of internal representation held by the operator may predict control behaviour. Although his findings are tentative they do suggest that different types of information are used in problem search and problem diagnosis. During search only the mental image (i.e. a mental picture of the plant) plays a role, whereas the mental model (i.e. an understanding of the cause–effect relationships between plant components) plays a more important role in diagnosis. Landeweerd explains that this is because search behaviour is working from symptoms to causes, whilst diagnosis relates the results from the search activities to probable effects. However, the correlations between the mental image and mental model data obtained by Landeweerd were not very high, and the internal representations may be moderated by other variables, such as learning or cognitive style.

A number of studies have suggested that the type of knowledge acquired during problem solving may indicate success in dealing with failures. In a comparison of training principles with procedures, the results indicate that rule-based reasoning is better for routine failures, whereas knowledge-based reasoning is better for novel situations (Mann and Hammer, 1986; Morris and

Rouse, 1985). Rouse and Rouse (1982) suggest that selection of strategies for problem solving tasks could be based upon cognitive style as certain styles may reflect more efficient behaviour. However, the results of further work indicate that the variations found in individuals highlight the need for more flexible training programmes.

In an analysis of the convergence or divergence of hypothesis testing in problem solving, Boreham (1985), suggests that success may be enhanced by the subject considering more hypotheses than absolutely required. This suggestion implies that a certain redundancy in options available may aid the task of problem solving by getting the subject to consider the problem further in order to justify their choice of intervention strategy. However, Su and Govindaraj (1986) suggest that the generation of a large set of plausible hypotheses actually degrades performance due to the inherent limitations of information processing ability. Providing many possible alternatives, therefore, makes the identification of the correct alternative more difficult, whereas a limited selection would presumably make the decision task easier.

Brehmer (1987) proposes that the increasing complexity of system dynamics makes the task of fault management more one of utilizing diagnostic judgment in a situation of uncertainty and less one of troubleshooting. The supervisory control task is becoming more like that of a clinician in diagnosing various states of uncertainty rather than the application of troubleshooting methods such as split-half strategies. Research on the diagnostic process suggests that the form of judgment tends to be simple (little information used, and it tends to be used in an additive rather than configurational way); the process is generally inconsistent, there are wide individual differences and individuals are not very good at describing how they arrived at judgments (Brehmer, 1987).

> The problem of fault diagnosis in complex systems arrives not from major catastrophic faults, but from cascades of minor faults that together overwhelm the operator, even though none would do so singly.
>
> Moray and Rotenburg (1989)

Thus the nature of the process plant may be considered to be greater than the sum of its parts due to the: inter-relation of the parts of the process plant, the system dynamics, many feedback loops and the inherent ambiguity of the information for diagnostic evaluation (Moray, 1980). This change in the nature of the task from troubleshooting to diagnostic judgement in a situation of uncertainty has implications for the way in which information is presented. As Goodstein (1985) suggests, this needs to change also. Goodstein proposes that the information should move away from the traditional physical representation of plant components toward a functional representation as, he suggests, this is closer to the operators' understanding of the plant. Thus the functional representation requires less internal manipulation.

Moray and Rotenberg's (1989) investigation into fault management in process control supported the notion that humans inherently prefer to deal with faults serially, rather than by switching between problems. They claim that this has serious implications for fault management in large complex systems, where any response to faults occurring late in the sequence of events would be greatly delayed, even if the later faults were of a higher priority than the earlier faults. It has been further proposed that in dealing with complex systems, humans are susceptible to certain 'primary mistakes'. These include: an insufficient consideration of processes in time, difficulties in dealing with exponential events and thinking in terms of causal series rather than causal nets (Reason, 1988c). These factors combined may help explain why the operators' understanding of the system state may not always coincide with the actual system state (Woods, 1988). Clearly the investigative task is very complex, and a means of representation to aid the operators' activities needs to consider the points mentioned here.

Correct

Corrective actions are those actions that result from the previous cognitive modes in response to the alarm(s). In a field study, Kortland and Kragt (1980), found that the limited number of actions that followed an alarm signal suggested that the main functions of the annunciator system under examination were to be found in its usefulness for monitoring. This supports Moray and Rotenberg's (1989) assertions that low observable physical activity is not necessarily accompanied by low mental activity. The majority of signals analysed by Kortland and Kragt (1980) were not actually 'alarms' in the sense that a dangerous situation was likely to occur if the operator did not intervene, and this must have led to its use as a monitoring tool, which has also been observed in other studies (Kragt and Bonten, 1983). However, they found that during periods of high activity the operator may pay less attention to individual signals, and mistaken actions could occur. Thus, lapses in attention in early AIA modes may lead to inappropriate corrective actions. The choice of compensatory actions is made by predicting the outcome of the alternatives available, but these evaluations are likely to be made under conditions of high uncertainty (Bainbridge, 1984). Bainbridge offers eight possible reasons for this uncertainty in the operator:

- action had unpredictable or risky effects;
- inadequate information about the current state of the system;
- wrong assumption that another operator had made the correct actions;
- precise timing and size of effects could not be predicted;
- no knowledge of conditions under which some actions should not be used;
- no knowledge of some cause–effect chains in the plant;
- difficulty in assessing the appropriateness of his/her actions;
- distractions or preoccupations;

It is assumed that knowledge embodied in the form of a coherent representation of the system and its dynamics (i.e. a conceptual model) would facilitate control actions, but the evidence is not unequivocal (Duff, 1989). Reason (1988a) suggests, in an analysis of the Chernobyl incident, that plant operators operate the plant by 'process feel' rather than a knowledge of reactor physics. He concludes that their limited understanding was a contributing factor in the disaster. However, under normal operation the plant had given service for over three decades without major incident. It was only when their actions entered into high degrees of uncertainty (as listed by Bainbridge, 1984) and combined with other 'system pathogens' that disaster became inevitable (Reason, 1988a).

Open-loop control strategies appear to be preferable in process control because of the typically long time constants between an action being taken and the effect of that manipulation showing on the display panel. Under such circumstances, closed-loop process manipulation might be an inefficient and potentially unstable strategy (Wickens, 1984). Under consideration of the 'multiple resources' representation of information processing, Wickens (1984) proposes that 'stimulus–cognitive processing–response' (SCR) compatibility will enhance performance, and conversely 'SCR' incompatibly would be detrimental to performance. This relationship means that the alarm display needs to be compatible with the response required of the operator. This framework may be used to propose the hypothetical relationship between alarm type and compatible response. This may be summarized as: text and speech based alarms would require a vocal response, whereas mimic and tone based alarms would require a manual response. Annunciator alarms appear to have both a spatial and a verbal element. Presumably they could, therefore, allow for either a verbal or a manual response. This last example highlights some difficulties with the SCR compatibility idea. Firstly, just because an input modality appears to be either verbal or spatial it does not necessarily allow for a simple classification into an information processing code. Secondly, many real life situations cross both classifications. Thirdly, control rooms usually require some form of manual input, and speech based control rooms, although becoming technically feasible, may be inappropriate for some situations (Baber, 1991a). Finally, Farah (1989) has indicated that recent research suggests that the distinction between information processing codes may not be as clear as the multiple resource theorists believe.

Rouse (1983) argues that diagnosis and compensation are two separate activities that compete with each other. The AIA model presents investigation and correction as separate stages, but the second activity may be highly dependent upon the success of the first. However, Rouse (1983) suggests that concentrating on one of the activities to the exclusion of all others may also have negative consequences. Therefore, whilst the two activities are inter-dependent, they have the potential for being conflicting, and Rouse asserts that this underlies the potential complexity of dealing with problem solving at multiple levels.

It is important to note that the presence of the alarm by itself may not directly suggest what course of action is required. An alarm only reports that a particular threshold has been crossed.

Monitor

Assessing the outcome of one's actions in relation to the AIAs can be presumed to be the monitor stage. It may appear to be very similar to the analyse stage in many respects, as it may involve an information search and retrieval task. Essentially, however, this mode is supposed to convey an evaluation of the effect of the corrective responses. Baber (1990) identifies three levels of feedback an operator may receive in control room tasks, these are:

- reactive;
- instrumental
- operational.

Reactive feedback may be inherent to the device, (for example, tactile feedback from a keyboard) and is characteristically immediate. Instrumental feedback relates to the lower aspects of the task, such as the typing of a command returning the corresponding message on the screen. Whereas operational feedback relates to higher aspects of the task, such as the decision to send a command which will return the information requested. These three types of feedback can be identified on a number of dimensions (Baber, 1990):

- temporal aspects;
- qualitative information content;
- relative to stage of human action cycle.

The temporal aspects refer to the relation in time for the type of feedback. Obviously reactive is first and operational is last. The content of the information relates to the degree of 'task closure' (Miller, 1968) and ultimately to a model of human action (Norman, 1986). Much of the process operator's behaviour may appear to be open-loop and therefore does not require feedback. This open-loop behaviour is due to the inherent time lag of most process systems. The literature shows that if feedback is necessary for the task, delaying the feedback can significantly impair performance (Welford, 1968). Therefore under conditions of time lag, the process operator is forced to behave in an open-loop manner. However, it is likely that they do seek confirmation that their activities have ultimately brought the situation under control, so delayed operational feedback should serve to confirm their expectations. If confirmation is sought, there is a danger that powerful expectations could lead the operator to read a 'normal' value when an 'abnormal' value is present (Moray and Rotenberg, 1989).

The operator will be receiving different types of feedback at different points in the AIAs. In the accept and correct stages they will get reactive and instrumental feedback, whereas in the monitor stage they will eventually get operational feedback. The operator is unlikely to have difficulties in interpreting and understanding reactive and instrumental feedback, if it is present, but the same is not necessarily true of operational feedback. The data presented to the operator in terms of values relating to plant items such as valves, pumps, heaters, etc., may be just as cryptic in the monitor stage as when they were requested in the investigative stage. Again the operator may be required to undertake some internal manipulation of this data in order to evaluate the effectiveness of his corrective actions, which may add substantially to the operator's mental workload.

The monitoring behaviour exhibited by humans is not continuous, but is characterized by intermittent sampling. As time passes, the process operator will become less certain about the state of the system. Crossman, Cooke *et al.* (1974) attempt to show this as a 'probability times penalty' function, where probability refers to the subjective likelihood of a process being out of specification and penalty refers to the consequences. This is balanced against the cost of sampling which means that attention will have to be diverted away from some other activity. They suggest that when payoff is in favour of sampling, the operator will attend to the process, and as soon as the uncertainty is reduced, attention will be turned to the other activities. However, they point out that monitoring behaviour is also likely to be influenced by other factors, such as: system dynamics, control actions, state changes, and the operator experienced memory decay. For example the processes may drift in an unpredictable way; operators might not know the precise effects of a control action; the process plant might be near its operational thresholds; more experienced operators might typically sample less frequently than novices, and if the operators forget values or states they might need to resample data. Crossman, Cooke *et al.* (1974) conclude from their studies that to support human monitoring of automatic systems, the system design should incorporate: a need for minimal sampling, a form of guiding the operator's activities to minimize workload, and enhanced display design to optimize upon limited attentional resources.

Conclusions

Activity in the control room may be coarsely divided into two types: routine and incident. This chapter has only considered the alarm handling aspects of the task, which have been shown to cover both routine and incident activities. However, the incident handling activities take only a small part of the operator's time, approximately 10 per cent (Baber, 1990; Rienhartz and Rienhartz, 1989) and yet they are arguably the most important part of the task. A generic structure of the task would be:

- information search and retrieval;
- data manipulation;
- control actions.
  (from: Baber, 1990)

This highlights the need to present the information to the operator in a manner that always aids these activities. Firstly, the relevant information needs to be made available to the operator to reduce the search task. The presence of too much information may be as detrimental to task performance as too little. Secondly, the information should be presented in a form that reduces the amount of internal manipulation the operator is required to do. Finally, the corrective action the operator is required to take should become apparent from both the second activity and the control interface, i.e. they can convert intention into action with the minimum of interference.

It seems likely that the requirements from the alarm system may be different in each of the six stages. For example:

- conspicuity is required in the observation stage;
- time to identify and acknowledge is required in the acceptance stage;
- information to classify with related context is required in the analysis stage;
- underlying cause(s) need to be highlighted in the investigation stage;
- appropriate corrective action afforded is required in the correction stage; and
- operational feedback is required in the monitoring stage.

Therefore, it appears that alarm information should be designed specifically to support each of the stages in the alarm initiated activities (AIA) model. The difficulty arises from the conflicting nature of the stages in the model, and the true nature of alarms in control rooms, i.e. they are not single events occurring independently of each other but they are related, context-dependent and part of a larger information system. Adding to this difficulty is the range of individual differences exhibited by operators (Marshall and Shepherd, 1977) and there may be many paths to success (Gilmore, Gertman *et al.*, 1989). Therefore, a flexible information presentation system would seem to hold promise for this type of environment.

The model of AIAs (Figure 6.1) is proposed as a framework for research and development. Each of the possible alarm media has inherent qualities that make it possible to propose the particular stage of the AIA it is most suited to support. Therefore, it is suggested that speech favours semantic classification, text lists favour temporal tasks, mimics favour spatial tasks, annunciators favour pattern matching tasks and tones favour attraction and simple classification. Obviously a combination of types of information presentation could support a wider range of AIAs, such as tones and text together. These are only working hypotheses at present and more research needs to be undertaken in the AIAs to arrive at preliminary conclusions. It is proposed that:

1. the 'observe' stage could benefit from research in detection and applied vigilance;
2. 'accept' could benefit from work on group versus single acknowledgement;
3. 'analyse' could benefit from work on classification and decision making;
4. 'investigate' requires work from problem solving and diagnosis;

5. 'correct' needs work on affordance and compatibility; and
6. 'monitor' needs work on operational feedback.

However, it is already proposed that the best method of presenting alarm information will be dependent upon what the operator is required to do with the information and on the stage of AIA model the information is used. Therefore the alarm types need to be considered in terms of the AIA. This may be undertaken through a systematic comparison of combinations of alarm message across task types to investigate empirically the effect of messages type and content on performance.

In summary, it is proposed that the alarm system should support the AIA. Observation may be supported by drawing the operators' attention, but not at the expense of more important activities. Acceptance may be supported by allowing the operators to see which alarm they have accepted. Analysis may be supported by indicating to the operators what they should do next. Investigation may be supported by aiding the operators in choosing an appropriate strategy. Correction may be supported through compatibility between the task and the response. Finally, monitoring may be supported by the provision of operational feedback. The design of alarm information needs to reflect AIA, because the purpose of an alarm should not be to shock operators into acting, but to get them to act in the right way.

References

Baber, C., 1990, 'The human factors of automatic speech recognition in control rooms, unpublished PhD thesis, Aston University, Birmingham.

Baber, C., 1991a, *Speech technology in control room systems: a human factors perspective*, Chichester: Ellis Horwood.

Baber, C., 1991b, Why is speech synthesis inappropriate for control room applications? In Lovesey, E.J. (Ed.) *Contemporary Ergonomics: Ergonomics Design for Performance*, London: Taylor & Francis.

Baber, C., Stammers, R.B. and Taylor, R.T., 1990, Feedback requirements for automatic speech recognition in control room systems, in Diaper, D., Gilmore, D., Cockton, G. and Shackel, B. (Eds) *Human-Computer Interaction: INTERACT '90*, pp. 761–6, Amsterdam: North-Holland.

Bainbridge, L., 1983, The ironies of automation, *Automatica*, **19** (6), 775–9.

Bainbridge, L., 1984, Diagnostic skill in process operation, *International Conference on Occupational Ergonomics*, 7–9 May, Toronto.

Boreham, N.C., 1985, Transfer of training in the generation of diagnostic hypotheses: the effect of lowering fidelity of simulation, *British Journal of Education Psychology*, **55**, 213–23.

Brehmer, B., 1987, Models of diagnostic judgements, in Rasmussen, J., Duncan, K. and Leplat, J. (Eds) *New Technology & Human Error*, Chichester: Wiley.

Broadbent, D., 1990, Modelling complex thinking, *The Psychologist*, **3** (2), 56.

Coury, B.G. and Pietras, C.M., 1989, Alphanumeric and graphical displays for dynamic process monitoring and control, *Ergonomics*, **32** (11), 1373–89.

Crossman, E.R.F.W. and Cooke, J.E., 1974, Manual control of slow response systems, in Edwards, E. and Lees, F.P. (Eds) *The Human Operator in Process Control*, London: Taylor & Francis.

Crossman, E.R.F.W., Cooke, J.E. and Beishon, R.J., 1974, Visual attention and the sampling of displayed information in process control, in Edwards, E. and Lees, F.P. (Eds) *The Human Operator in Process Control*. London: Taylor & Francis.

Duff, S.C., 1989, Reduction of action uncertainty in process control systems: the role of device knowledge, in *Contemporary Ergonomics 1989*, Proceedings of the Ergonomics Society 1989 Annual Conference; 3–7 April, London: Taylor & Francis.

Easterby, R., 1984, Tasks, processes and display design, in Easterby, R. and Zwaga, H. (Eds) *Information Design*, Chichester: Wiley.

Edwards, E. and Lees, F.P., 1974, *The Human Operator in Process Control*, London: Taylor & Francis.

Edworthy, J. and Loxley, S., 1990, Auditory warning design: the ergonomics of perceived urgency, in Lovesey, E.J. (Ed.) *Contemporary Ergonomics 1990: Ergonomics setting the standards for the 90s*; 3–6 April, pp. 384–8, London: Taylor & Francis.

Elkind, J.I., Card, S.K., Hochberg, J. and Huey, B.M., 1990, (Eds) *Human Performance Models for Computer-Aided Engineering*, Boston: Academic Press;

Eysenck, M.W., 1984, *A Handbook of Cognitive Psychology*, London: Lawrence Erlbaum Associates.

Farah, M.J., 1989, Knowledge from text and pictures: a neuropsychological perspective, in Mandl, H. and Levin, J.R. (Eds) *Knowledge Aquisition from Text and Pictures*, pp. 59–71, North Holland: Elsevier.

Gilmore, W.E., Gertman, D.I. and Blackman, H.S., 1989, *User-Computer Interface in Process Control*, Boston: Academic Press.

Goodstein, L.P., 1985, *Functional Alarming and Information Retrieval*, Denmark: Risø National Laboratory, August Risø-M-2511. 18.

Goodstein, L.P., Andersen, H.B. and Olsen, S.E., 1988, *Tasks, Errors and Mental Models*, London: Taylor & Francis.

Hale, A.R. and Glendon, A.I., 1987, *Individual Behaviour in the Control of Danger*, Amsterdam: Elsevier.

Hoc, J-M., 1988, *Cognitive Psychology of Planning*, London: Academic Press.

Hockey, G.R.J., Briner, R.B., Tattersall, A.J. and Wiethoff, M., 1989, Assessing the impact of computer workload on operator stress: the role of system controllability, *Ergonomics*, **32** (11), 1401–18.

Jenkinson, J., 1985, *Alarm System Design Guidelines*, Central Electricity Generating Board, September, GDCD/CIDOCS 0625.

Johannsen, G., 1988, Categories of human operator behaviour in fault management situations, in Goodstein, L.P., Andersen, H.B. and Olsen, S.E. (Eds) *Tasks, Errors and Mental Models*, pp. 251–58, London: Taylor & Francis.

Kortlandt, D. and Kragt, H., 1980, Process alarm systems as a monitoring tool for the operator, in *Proceedings of the 3rd International Symposium on Loss Prevention and Safety Promotion in the Process Industries.*; September 15–19, Basle, Switzerland, Vol. 1, pp. 10/804–10/814.

Kragt, H. and Bonten, J., 1983, Evaluation of a conventional process alarm system in a fertilizer plant. *IEEE Transactions on Systems, Man and Cybernetics*, **13** (4), 589–600.

Landeweerd, J.A., 1979, Internal representation of a process, fault diagnosis and fault correction, *Ergonomics*, **22** (12), 1343–51.

Mann, T.L. and Hammer, J.M., 1986, Analysis of user procedural compliance in controlling a simulated process, *IEEE Transactions on Systems, Man and Cybernetics*; **16** (4), 505–10.

Marshall, E. and Shepherd, A., 1977, Strategies adopted by operators when diagnosing plant failures from a simulated control panel, in *Human Operators and Simulation*, pp. 59–65. London: Institute of Measurement & Control.

Miller, R.B., 1968, Response time in man–Computer conversational Transactions, *Proceedings of the Spring Joint Computer Conference*, **33**; 409–21, Reston, Virginia: AFIRS Press.

Molitor, S., Ballstaedt, S-P. and Mandl, H., 1989, Problems in knowledge acquisition from text and pictures, in Mandl, H. and Levin, J.R. (Eds) *Knowledge Acquisition from Text and Pictures*, pp. 3–35, North Holland: Elsevier.

Moran, T.P., 1981, An applied psychology of the user, *Computer Surveys*, **13** (1), 1–11.

Moray, N., 1980, The role of attention in the detection of errors and the diagnosis of failures in man–machine systems, in Rasmussen, J. and Rouse, W.B. (Eds) *Human Detection and Diagnosis of System Failures*, New York: Plenum Press.

Moray, N. and Rotenberg, I., 1989, Fault management in process control: eye movements and action, *Ergonomics*; **32** (11), 1319–42.

Morris, N.M. and Rouse, W.B., 1985, The effects of type of knowledge upon human problem solving in a process control task, *IEEE Transactions on Systems, Man and Cybernetics*, **15** (6), 698–707.

Newell, A. and Simon, H.A., 1972, *Human Problem Solving*, Englewood Cliffs, NJ: Prentice Hall.

Norman, D.A., 1986, Cognitive engineering, in Norman, D.A. and Draper, S.W. (Eds) *User Centred System Design, Hillsdale*, NJ: Lawrence Erlbaum Associates.

Pal, J.K. and Purkayastha, P., 1985, Advanced man–machine interface design for a petroleum refinery plant, in Johannsen, G., Mancini, G. and Martensson, L. (Eds) *Analysis, Design and Evaluation of Man–Machine Systems*, pp. 331–37, Italy: Commission of the European Communities.

Perrow, C., 1984, *Normal accidents: Living with high risk technology*, New York: Basic Books.

Pew, R.W. and Baron, S., 1982, Perspectives on human performance modelling, in Johannsen, G. and Rijnsdorp, J.E. (Eds) *Analysis, Design & Evaluation of Man–Machine Systems*, Duesseldorf: IFAC.

Pew, R.W., Miller, D.C. and Freehrer, C.E., 1982, Evaluating nuclear control room improvements through analysis of critical operator decisions, *Proceedings of the Human Factors Society 25th Annual Meeting*, pp. 100–4.

Rasmussen, J., 1976, Outlines of a hybrid model of the process plant operator, in Sheriden, T.B. and Johannsen, G. (Eds) *Monitoring Behaviour and Supervisory Control*, New York: Plenum Press.

Rasmussen, J., 1983, Skills, rules and knowledge; signals, signs and symbols, and other distinctions in human performance models, *IEEE Transactions on Systems, Man and Cybernetics*, **13** (3).

Rasmussen, J., 1984, Strategies for state identification and diagnosis in supervisory control tasks, and design of computer based support systems, in Rouse, W.B. (Ed.) *Advances in Man–Machine Systems Research*, pp. 139–93.

Rasmussen, J., 1986, Information processing and human–machine interaction, *An Approach to Cognitive Engineering*, North-Holland: Amsterdam.

Reason, J., 1988a, The Chernobyl errors, *Bulletin of the British Psychological Society*, **40**, 201–6.

Reason, J., 1988b, Framework models of human performance and error: a consumer guide, in Goodstein, L.P., Andersen, H.B. and Olsen, S.E. (Eds) *Tasks, Errors and Mental Models*, London: Taylor & Francis.

Reason, J., 1988c, Cognitive aids in process environments: prostheses or tools? In Hollnagel, E., Mancini, G. and Woods, D.D. (Eds) *Cognitive Engineering in Complex Dynamic Worlds*, pp. 7–14.

Reason, J., 1990, *Human Error* Cambridge: Cambridge University Press.

Reed, J. and Kirwan, B., 1991, An assessment of alarm handling operations in a central control room, in Quéinnec, Y. and Daniello, F. (Eds) *Designing for Everyone*, London: Taylor & Francis.

Reinartz, S.J. and Reinartz, G., 1989, Analysis of team behaviour during simulated nuclear power plant incidents, In Megaw, E.D. (Ed.) *Contemporary Ergonomics 1989*, Proceedings of the Ergonomics Society 1989 Annual Conference 3–7 April, pp. 188–93, London: Taylor & Francis.

Rouse, W.B., 1983, Models of human problem solving, *Automatica*, **19**, 613–25.

Rouse, S.H. and Rouse, W.B., 1982, Cognitive style as a correlate of human performance in fault diagnosis tasks, *IEEE* Transactions on Systems, Man and Cybernetics, **12** (5), 649–52.

Singleton, W.T., 1989, *The Mind at Work*, Cambridge: Cambridge University Press.

Sorkin, R.D., 1989, Why are people turning off our alarms? *Human Factors Bulletin*, **32**, 3–4.

Stanton, N.A., 1992, 'Human factors aspects of alarms in human supervisory control tasks,' unpublished Phd thesis, Aston University: Birmingham.

Stanton, N.A., 1993, Operators reactions to alarms: fundamental similarities and situational differences, *Proceedings of the Conference on Human Factors in Nuclear Safety*, Le Meridien Hotel, London, 22–23 April.

Stanton, N.A. and Booth, R.T., 1990, The psychology of alarms. In Lovesey, E.J. (Ed.) *Contemporary Ergonomics*, London: Taylor & Francis.

Stanton, N.A., Booth, R.T. and Stammers, R.B., 1992, Alarms in human supervisory control: a human factors perspective, *International Journal of Computer Integrated Manufacturing*, **5** (2), 81–93.

Su, Y-L. and Govindaraj, T., 1986, Fault diagnosis in a large dynamic system: experiments on a training simulator, *IEEE Transactions on Systems, Man and Cybernetics*, **16** (1), 129–41.

Swain, A.D. and Weston, L.M., 1988, An approach to the diagnosis and misdiagnosis of abnormal conditions in post-accident sequences in complex man-machine systems, in Goodstein, L.P., Andersen, H.B. and Olsen, S.E. (Eds) *Tasks, Errors and Mental Models*, London: Taylor & Francis.

Welford, A.T., 1968, *Fundementals of Skill*, London: Methuen.

Wickens, C.D., 1984, *Engineering Psychology and Human Performance*, Columbus, Ohio: Merrill.

Wickens, C.D. and Kessel, C., 1981, Failure detection in dynamic systems, in Rasmussen J. and Rouse, W.B. (Eds) *Human Detection and Diagnosis of Systems Failures*, New York: Plenum Press, pp. 155–69.

Woods, D.D., 1988, Coping with complexity: the psychology of human behaviour in complex systems, in Goodstein, L.P., Andersen, H.B. and Olsen, S.E. (Eds) *Tasks, Errors and Mental Models*, pp. 128–48, London: Taylor & Francis.

7

Supervisory control behaviour and the implementation of alarms in process control

H.J.G. Zwaga and H.C.M. Hoonhout

Introduction

Distributed control systems (DCSs) with fully VDU-based operations consoles were introduced in the petrochemical industry in the mid-seventies. Currently, all new plants, refurbished plants, new and refurbished production platforms, etc. are equipped with centralized VDU-based distributed control systems to support the operator's supervisory control task.

DCSs manufactured by companies such as Foxboro, Honeywell and Yokogawa are acquired by the petrochemical industry almost as a standard piece of equipment for the presentation of process information and for control of the process. The safeguarding system is, however, always independent of the DCS. By comparison, the introduction of centralized control systems with VDU-based operations consoles in the nuclear power industry has been a slower process. Certainly, more risks are involved in this industry, although here too, control and safeguarding are separated. Other factors that probably determine the slower introduction of DCSs are the usually large size of the processes and the reluctant introduction of new technology. The many rules and regulations that authorities have ordained in the nuclear power industry necessitate a conservative approach in adopting new technology.

Whether the fast introduction of DCSs in the petrochemical industry was possible because the consequences of an accident were considered to be less serious (or, at least, to attract less public attention), or because processes were less complicated, should be the subject of debate elsewhere. A fact is that DCSs have been installed, and still are, on a regular basis, and there have been no serious accidents officially attributed to the DCS and its VDU-based man–machine interface. It should be stressed again here, that a DCS is the first layer in controlling the plant. Independent safeguarding equipment will cope with DCSs failure, including human operator failure.

Here, we will describe and discuss some aspects of human factors requirements and human factors involvement in the implementation of DCSs in various processes of one large petrochemical company. From publications on plant operation in other companies in the petrochemical process industry it can be concluded that our experiences are not atypical (Kortlandt and Kragt, 1980; Bainbridge, 1987; van der Schaaf, 1989).

The two main themes discussed in this paper will be: 1) the characteristics and aims of supervisory monitoring and control behaviour of operators, and 2) the effects of alarm implementation and alarm information facilities on operator behaviour, performance and appreciation of a DCS. It will be shown that the operator's supervisory control behaviour determines, to a large extent, the requirements with regard to alarm implementation and alarm information facilities.

The introduction of distributed control systems

Starting in the mid-seventies, DCSs have been introduced at an increasing rate in the petrochemical industry. There are several reasons for this: energy conservation, minimization of off-spec production, integration of process units to improve efficiency, and the need of management to obtain better and more accurate and detailed process information. Distributed control systems provide the technology for these requirements.

The changes in control room instrumentation, i.e. from panel mounted displays and controls to VDU-based consoles, concurrent with a decrease in the number of operators and an increase in instrumentation density, have considerably increased the operator's workload and responsibility. But not only workload and responsibility have increased, a higher level of skill is also required, because disturbances no longer tend to occur in isolation. Due to plant integration and because intermediate storage is no longer available, the effect of disturbances is much wider and can spread rapidly over more parts of the process.

The extent to which a VDU-based DCS console could have an impact on the operator's task was not fully realized by system designers and instrument engineers. For them it was a major step forward from the single automatic control loop to a DCS. The crucial question was whether an operator could perform his task with predominantly sequentially, rather than simultaneously, available information and control facilities. The straightforward mapping of the plant with dedicated information sources and controls was replaced by 'general purpose' CRT screens and controls. With a DCS, choices had to be made with regard to the most important parameters to be presented in a display, to the rendition of information, the combination of different information sources, and ways of access to information. These choices were not required when manufacturers were marketing nothing more complicated than the conventional panel-mounted instruments such as dials, switches and recorder controllers. Information about the way operators perform their task suddenly became important. This information would determine the facilities that had to be available in the DCS operator interface and, more specifically, it would show whether a DCS was feasible at all, given the hardware functionality available at that time, i.e. the beginning of the seventies. Even in the mid-eighties the system response time for updated mimics was usually 5–10 seconds (e.g. for the Honeywell TDC 2000), restricting their usefulness considerably.

DCS design philosophy

The change in appearance of the man–machine interface was so substantial that this choice required justification. The well documented engineering-oriented design 'philosophy' of Dallimonti is used to justify this step from conventional instrumentation to DCS. In two publications Dallimonti (1972, 1973) described the way operators perform their task and how the facilities in the operator interface of the Honeywell TDC 2000 supported this. From the results of a number of field evaluations, conducted prior to the conceptual design of the Honeywell TDC 2000, Dallimonti concluded that operators perform their task according to the 'management-by-exception' approach. As a rule operators are triggered into action by an upcoming alarm. Analysis of the disturbed process conditions then moves from the use of global to increasingly more detailed levels of information, resulting in a diagnosis and remedial control action. Alarm information and general qualitative information about the status of the process are the prime sources of information monitored by an operator on an ongoing basis.

Dallimonti stated that displays in a hierarchical ordering best suit the operator's needs. Based on his analysis, he also defined the information content and presentation of the different display types and how to get access to them. He also concluded that there is only a limited need for simultaneously presented process information. Using one, two, or three VDUs is sufficient for the predominantly sequentially presented information. Dallimonti (1972) summarized his conclusions as follows:

- operation by exception is pretty much the way operators monitor, whether they consciously recognize it or not;
- graphic panels and other large mimic displays are of questionable value after the initial learning period;
- at the first level of process monitoring, operators do not use quantitative information;
- the continuing trend to centralized control is resulting in reduced man-

Table 7.1   Overview of the evaluations conducted

Site	Size (loops)	Instrumentation	Period
Plant-1	170	Honeywell TDC-2000 (basic rel. 315)	1983–84
Plant-2	160	Honeywell TDC-2000 (basic rel. 315) + supervisory station	1985
Plant-3	170	Honeywell TDC-2000 (enhanced rel. 520)	1986
Plant-4	145	Foxboro Spectrum (Videospec IV + Fox-300)	1986
Plant-5	50	Honeywell TDC-2000 (enhanced rel. 520)	1986
Plant-6	150	Siemens Teleperm M system	1990
3 Production platforms	50–60	Honeywell TDC-2000, Honeywell TDC-3000, Fischer Provox	1991

power with a resultant increase in the number of supervised loops per operator – and it is working.

The description of the operator's task, based on Dallimonti's field evaluations, apparently fitted in well with the opinions and experiences of instrument engineers and other manufacturers such as Foxboro (Fraser, 1978). Hartmann and Braun (Flippance and Shaw, 1978) followed rapidly with their systems based on the same 'philosophy'. Given the technological advantages of a DCS over conventional instrumentation, many project teams (i.e. instrument engineers) were eager to select a DCS as *the* control system for their process.

However, from the moment the first few systems were operational, complaints emerged about difficulties in the use of the interface, especially about keeping a proper overview of the system's status, and about the need to work with more than one panel operator. Complaints started filtering through that operators did not use these systems in the way described by the manufacturers. These complaints were in line with the results of studies of hybrid systems (operator interfaces with conventional instrumentation combined with VDU-presented information). Kortlandt and Kragt (1980), and Zwaga and Veldkamp (1984) had already suggested that operators often did not work according to the management-by-exception principle as assumed by the system designers.

Evaluation method

We conducted a number of field studies in the period 1984–1991 in order to evaluate more objectively the prevailing assumption at the time about the way operators performed their tasks, and to determine how they actually used the DCS interface. Table 7.1 presents an overview of the systems and the size of the processes. All systems had been operational for 18 months or longer at the time of the study.

Procedure

Our method consisted of two parts. The first part involved the systematic recording of operator activities during an observation period, and the second part consisted of structured interviews with operators from each shift.

The evaluation started with an observation period in the control room lasting 60 to 90 hours. Over this period operator activities were recorded during all parts of the day and throughout all shifts. The observations were completed in four or five days. The activities were recorded continuously for periods of six hours by a team of two observers. Usually, there were two or three teams of observers. During this observation period the following data were recorded:

- the frequency and duration of use of the available DCS information facilities;
- other activities of the panel operator, such as administrative work, communication, etc.;
- situational information, such as mode changes, on-coming alarms, disturbances, number of operators at the console etc.

For the recording purposes, all activities of an operator were coded and defined (e.g. writing a report and completing a log sheet were both coded as 'admin'). All types of displays available in the DCS were also coded, and the observer on duty recorded what type of display was requested on each of the VDU-screens. Finally, the position at the console of up to three operators was recorded. A personal computer was used for data recording and storage. Alarm data, available from the DCS alarm printer, were merged with the observational data at a later stage.

The aim of the first part of the evaluation procedure was to obtain objective information about the operator's way of working and about the extent to which his/her actions depended on the process and the facilities provided by the DCS.

The second part of the evaluation consisted of individually conducted structured interviews (60–90 minutes) with two to three panel operators from each shift and at least two shift supervisors.

The topics covered in all interviews were:

- work procedure during quiet and upset process conditions;
- performance of the alarm system;
- specific use of the different display types;
- the appreciation of the introductory training programme for the new DCS interface.

The interviews were conducted during the second half of the observation period. This permitted the formulation of questions to specifically clarify difficult-to-interpret aspects of the supervisory behaviour observed. Furthermore, it was easier to make a distinction between actual working behaviour

Table 7.2   Operator paging behaviour

Site	% paging
Plant-1	45
Plant-2	60
Plant-3	37

and the operators' opinion about their task. As a check on the usefulness of the interviews, a questionnaire was developed for the first evaluation study, covering the same subjects as in the interviews.

The results of the questionnaires (80 operators) compared very well with the results of the interviews. However, the questionnaire was dropped from the procedure and only interviews were retained in consecutive studies. The structured interviews were preferred over the questionnaire not only for reasons of cost-effectiveness, but also because the in-depth information from the interviews was considered to be more valuable than the information from the questionnaire. Further details about the method and more detailed results are presented in Zwaga and Veldkamp (1984) and Swaanenburg, Zwaga *et al.* (1988).

Results of the evaluation studies

Supervisory monitoring behaviour

The results of both the observations and the interviews clearly indicated that operators do not perform their task based on the management-by-exception principle. The studies listed in Table 7.1 all indicate that operators prefer to monitor the process quite intensely. They need information about the dynamic state of the process because their prime concern is to know, with an acceptable degree of certainty, that the process is not going to confront them with unpleasant surprises. To infer that the process is running smoothly, because there are no alarms, is not sufficient for them. The observation results show that they perform this monitoring (updating) task in such a way that it is efficiently tuned to the status of the process. If the process is stable, it is far less often checked than when it is not completely balanced. How they perform this task, and what kind of information they use, depends mainly on the size and complexity of the plant. Detailed analysis of this monitoring behaviour shows that operators mainly request information on groups of related variables (available as so-called group displays). Depending on the process conditions, a series of requested group information displays are mixed with requests for (mainly) trend displays and mimic displays. This 'paging' through the DCS displays has been analysed by defining three consecutive display requests with a request interval of 10 seconds or less, as a minimum 'paging string'. Table 7.2 shows the percentage of the display requests belonging to paging strings in the three main evaluation studies.

This 'paging' through the process and processing of the information requested, constitutes a heavy task load. Given this way of working during normal process conditions, it is not surprising that during abnormal process conditions, when alarms interfere with updating activities, an operator will quite soon ask for assistance to delegate parts of his task to a second operator.

The main reason that operators work as described above is that they, as well as operational staff, consider it the operator's task to *prevent* alarms, rather than to *react* to alarms. For this reason, information directly related to specific variables is needed (group, trend and mimic displays), because only this kind of information allows the operator to predict future states of the process and, if necessary, to take preventive action. The displays intended by system designers for the operator's supervisory task, i.e. the overview displays and alarm displays, are usually of little use for the operator during normal process conditions. In contrast to 'management-by-exception', it is 'management-by-awareness' that most of the time guides the operator's behaviour. In practice, as our observation studies have shown, an operator will supervise a process, or only part of a process, using mainly one or the other strategy. He will supervise a process, or parts of a process, with quickly changing conditions using the management-by-awareness strategy. Only if the stability of the process is high will the operator consider it sufficient to check just a few key variables occasionally, and he will rely mainly on the alarm system. Considering all observation studies, there is no doubt that, given the usually high alarm frequency, the prevailing strategy used is management-by-awareness.

This systematic monitoring of the dynamic state of the process by the operator was, as a mental activity, probably not apparent to Dallimonti, because he mainly focused on the physical activities of the operator and not on the information processing activities of the operator. Provision of the DCS interface has led to these monitoring activities becoming more observable, because a display has to be selected before the information can be scanned.

Operator behaviour and implications for interface design

In principle, a DCS makes information available in a serial way. This hinders the operator's use of the awareness strategy. Scanning and comparing information constitute the actual task. To do so, the DCS forces the operator to perform interfering tasks, i.e. consciously selecting displays, processing the information, integrating it with information already obtained, deciding which display to select next, etc.

A serious disadvantage of sequentially presented information is that operators show a strong tendency to make a decision to act as soon as the information is pointing in a certain direction. Reconsidering decisions in the light of later information appears to be a difficult step to make. Rasmussen and Vicente (1989) call this 'the principle of the point of no return'.

On many sites sets of pen recorders have been built into the consoles to compensate for this lack of information presented in parallel. Manufacturers offer console modules especially for this purpose. To lower the display request load, some sites have successfully installed co-ordinated work stations, that provide related information in parallel. Using an additional co-ordinated selection keyboard, one single request results in the presentation of not only a group display, but also of two additional displays with related information: a display with related trends, and a mimic display (Plant-4, Table 7.1). To achieve this, an operator's work place needs at least four screens: three for the displays already mentioned, and a screen for the alarm list display.

DCS manufacturers have also provided more effective facilities for the presentation of information in parallel. The more recent DCSs no longer have predefined display types. The displays can be built from combinations of different types of display features (groups, mimics, trends, etc.). This facility certainly is an advantage over the DCSs 'fixed' display types, because it can reduce the need to page through displays. One disadvantage, however, is that the use of this facility to present the operators with parallel information, requires a much more systematic and operation-oriented approach in the engineering phase than was needed with the DCSs with only standard displays. For each process condition the most effective operation strategy, i.e. management-by-awareness or management-by-exception, should be determined. Based on those decisions, display and control facilities have to be defined. The need for separate displays for supervisory updating and for disturbance analysis has to be considered. Overloading displays with information must be prevented. User needs and preferences, and procedures to design effective displays with regard to content as well as format, are discussed in detail by Hoonhout and Zwaga (1993) and Zwaga (1993).

Thus, now even more than before, the operators' task should be the starting point for decisions about the operator-process interface. Task analysis should be employed in the design of process control systems. The operator's information and control requirements under different process conditions should be determined objectively and the results should be used to design operations consoles, VDU-display formats and procedures to determine the information contents of displays.

This way of specifying displays and their use requires that operators should be trained in the use of the two operation strategies and the related functionality of the displays. It is evident that for this 'concept' training, a training course with a proper curriculum has to be developed.

User participation in system design

It is not just a coincidence that with the increase in flexibility of the recent types of DCSs, the involvement of operational staff in the engineering phase has also increased. In the petrochemical process industry, it is recognized that operators are an important source of information with regard to decisions about the usability of the system. At an increasing rate, it is decided to involve future users in the engineering phase of a project. Typically, future operators are involved in the design of the console and especially of the VDU-displays. Often, however, with the current types of DCSs allowing so much freedom of choice in display design, operator involvement seems to be a compensation for the lack of expertise on the part of the system design engineer.

It is evident, more so than with the fixed DCS display types, that the design of displays should be based on task requirements and operational experience. A disadvantage of this development, however, is that operators are not only used as a source of information, but they are also given the job of actually designing the DCS displays. This guarantees that the designs will be accepted by the operators. Whether those displays are indeed the best operational aids may often be questioned, however. They usually result in mimics crammed with information showing as close as possible replicas of PEFSs (Process and Engineering Flow Schemes) with an excessive and extensive use of colour. Although there are some exceptions it is significant that, from the very start of discussions about the operator interface, operators (and design engineers as well) have focused on the layout of the displays and not on their *purpose*, i.e. how the displays should support task performance by presenting the right kind of information, in the best possible way, and at the right moment.

The point is that being able to perform a task does not imply that one also knows how the task is performed. It is difficult to analyse one's own activities objectively. Operators who are quite able to perform their job, do not necessarily know *how* they do their job. However, task and job analysis techniques are not developed as just a scientific exercise. They are necessary because experience shows that simply asking people what they do and how they do it usually results in wrong conclusions. The conclusions of design engineers with respect to operator task performance hardly appear to be based on observations of, and discussions with, operators. Their conclusions seem to be based more on preconceived ideas about operator task performance, resulting in misconceptions such as: assuming the singular or at least dominant role of management-by-exception, ideas about the intrinsic usefulness of VDU-presented PEFSs, and the unrestrained use of colour and colour coding (an opinion which they share with most operators). Alty and Bergan (1992) are very outspoken about the need for human factors expertise in interface design and why design engineers are reluctant to make use of this expertise. In the context of the use of colour, they write:

> '. . . with an arrogance which seems to pervade the subject of interface design, the designers thought they knew best'.

Less openly, van Hal and Wendel (1992) discussed the involvement of operators in the design of VDU displays and stated that the contribution of operators should be judged with care, because they tend to adhere to existing task procedures. They point to the fact that it is difficult for operators to extrapolate from their current task to new ways of operating. Marshall (1992) presents an overview of the ways users can be involved in the design of computer based interfaces for power station control rooms. He warns that operator opinions will be biased by prior experience and by their interpretation of the functionality of a new process system. He points to the mediating role of human factors to resolve conflicts between designers and users, and implicitly emphasizes that the human factors specialist should have a prominent consulting role in decisions with regard to the application of information provided by the user.

Alarm information and alarm handling

DCS alarm presentation facilities and operator's task performance

Generalizing from all our observation studies and later studies using only structured interviews, it is evident that the alarm presentation facilities of a DCS are insufficient, whether implemented in a small or large refinery, in a chemical process, or on board a production platform. This is certainly the case during large disturbances or when more disturbances occur simultaneously.

For a single operator it is no problem to supervise a stable process, even a large one. The operators state that, in contrast, during a major disturbance or any other extensive off-normal condition, supervision of these same processes can hardly be done alone. The assistance of a second operator is needed to supervise the remaining undisturbed parts of the process; the second operator will also assist in the diagnosis of the disturbance and in the decisions on strategies for corrective action.

It is difficult to keep track of the progress of a disturbance with sequentially presented information. Alarm overview displays are only of some help, because they mostly provide an overview at too general a level, whereas the alarm list display is too detailed and lacks the functional order necessary to track the progression of the disturbance. The console manning data from the observation studies in Table 7.3 clearly show that for a substantial amount of time two operators work together at a console. During complex or extensive disturbances up to three operators are manning the console.

The prediction, made at the introduction of DCSs, that less operators would be needed in the control room, or that many more loops could be supervised by a single operator, has not come true. Many of the earlier designs force two operators to work simultaneously at a console intended for use by a single operator. This uncomfortable arrangement was a result of a lack of consideration of how the task would be performed.

Table 7.3 *Number of operators working at a console*

Site	Number of operators at the console (% of time)			
	0	1	2	3
Plant-1*	–	72	–	–	28	–
Plant-2	1	58	32	9
Plant-3	5	60	27	8
Plant-4	7	71	20	2

* In this study was only recorded whether one operator or more were working at the console

Table 7.4 *Alarm load in five control rooms*

Site	Average frequency (A/h)	Highest frequency (A/h)	% Oscillating Alarms
Plant-1	21	150	50
Plant-2	5	35	12
Plant-3	7	25	36
Plant-4	53	275	75
Plant-5	6	30	30

The data on alarm load show that alarms in the evaluated control rooms are by no means rare (Table 7.4). Even in a small plant with a process of medium complexity, an average number of alarms of six per hour was considered by the operators not to be excessive. Operators judged the average level in Plant-1 and Plant-4 high but not unacceptable. Operators probably had this opinion because many of the alarms in these plants originated from oscillating variables. On the one hand, these alarms added less to the alarm load because once an operator had decided that a series of oscillating alarms had started, he knew what to expect and could suffice with acknowledging the alarm. On the other hand, in the interviews the operators pointed out the dangers of this situation. Having to push an alarm-acknowledge button 1–5 times per minute, leaves an operator little time to do anything else. He might fail to notice changes in the oscillation because of an upcoming disturbance or, simultaneously appearing but non-related, alarms might be missed and acknowledge together with the oscillating alarms. In an early study of a fertiliser plant Kortlandt and Kragt (1980) had already stressed the need for some kind of alarm filtering or suppression, to reduce the number of alarms. Their data show that in the operator's opinion only 13 per cent of the alarms are important.

The results of the questionnaire, which was part of the evaluation procedure of Plant-1, clearly show that operators find the alarm system as a whole less useful during off-normal plant conditions than during stable process conditions (Figure 7.1). Even more drastic is their change in appreciation of

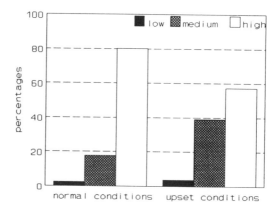

Figure 7.1   Alarm system appreciation during normal and upset conditions. Low, medium and high ratings in percentages of 80 operators.

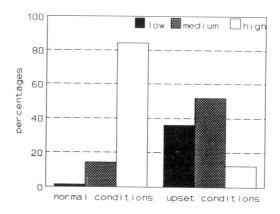

Figure 7.2   Overview display appreciation during normal and upset conditions. Low, medium and high ratings in percentages of 80 operators.

the help the overview display can provide during disturbed process conditions (Figure 7.2).

Reducing the alarm load resulting from oscillating alarms is possible, because only a few alarm points seriously suffer from this phenomenon. In addition, audits have shown that, given the ease with which alarms can be configured in a DCS (without any additional hardware costs), large numbers of alarms tend to be specified, because of the feeling of inherent safety it produces. The number of alarms has been further increased by the option to specify different levels of priority: some systems offer up to five levels. In practice, designers appear to restrict themselves to three levels. This, however, does not prevent large numbers from being implemented. Alarms are specified during the engineering phase with a strong involvement of each plant section's specialists, who understandably want to optimize the functioning of their part of the process, and therefore they want many alarm points in order to reach this goal. The proliferation of alarms is not only the result of every specialist requesting many alarms for his part of the process, but also because there are no proper criteria to decide which level of priority should be assigned to an alarm. Usually, the differences between priorities are based on characteristics of the action required to correct process conditions. Using descriptions for the different priority levels such as 'a small adjustment' or 'a fast action' allows opportunities for discussion when priorities have to be assigned. If a request for a high priority alarm is not granted, it might stand a better chance to get at least a medium priority level, etc.

It is necessary that design teams define more objective criteria for alarm priority levels in order to prevent an excessive increase in the number of alarms. Criteria based on the consequences for personnel, equipment and process might be a better approach. High priority alarms should be implemented to warn against immediate trips of the plant. They would then be related to life endangerment of personnel, environmental pollution and equipment in danger of destruction. Medium priority alarms would be related to production continuity, and low priority alarms to product quality.

Alarm information facilities and the operator's task requirements

As it appears now, there are a number of options to reduce the alarm load substantially. Proper attention to oscillating alarm points can give a substantial reduction of alarms, especially in the reduction of peak loads of alarms requiring no action. The use of objectively applicable criteria in the engineering phase to select alarm points and to assess their priority can prevent the implementation of superfluous alarms.

An important facility lacking in current DCSs is task-oriented alarm overview information. Even during an average disturbance, the alarm list displays are of little help. With more than one page of alarms (usually the maximum number of alarms on a page is about 20), it is difficult to keep track of the progression of a disturbance. Older operators refer in this respect to the help provided by the wall panels with conventional instruments and alarm lights, or the extensive alarm fascias. Once knowing their 'geography' it was possible to know the progression of a disturbance at a glance. Continuously presented alarm overview information to compensate for this lost overview, required by operators, was implemented in the earlier DCS control rooms in the form of hardwired alarm display panels (ADP). These ADPs have never been properly developed into a useful operational facility. No guidelines were developed for the information content and for the design of the panels. ADPs tend to disappear now, because they never had the support of design engineers. It is easy to prove that ADPs are costly, and they certainly can look like low-technology relics of a bygone era. All evidence, however, points to the need for this kind of alarm overview information. As an alternative, the functionality of an ADP could be built into one or two mimic VDU-displays. This, of course, requires a new way of presenting the information currently shown on an ADP. However, an ADP has advantages over VDU presentation. An ADP can be larger, so the information can be seen from a greater distance and, more importantly, by many people at the same time. This guarantees that the different members of the staff in the control room can quickly update themselves when the process conditions change and it facilitates discussion about the interpretation of the process conditions.

Building intelligence into the system interface is the next logical step. It can range from alarm filtering, e.g. suppression of derived alarms (i.e. alarms as a consequence of the first failure) to sophisticated decision support systems. Even the logically-evident step to suppress predictable alarms is not as straightforward as it seems. In all our studies, operators acknowledged that they were often distracted and irritated by the alarms following a first failure alarm, but they rejected the idea of suppressing those derived alarms. They want to keep these alarms, because they provide information about the correctness of their initial diagnosis, and also about the speed with which the disturbance is spreading. This provides extra information about the failure situation and sometimes even about the effect of their initial corrective actions.

With regard to the presentation of alarms, the possibility remains of having separate displays for first failure alarm and displays showing all alarms (e.g. an ADP) to allow alarms propagation analysis.

When even something relatively simple as alarm suppression can become questionable, because useful information might be lost by this suppression, the implementation of decision support systems will be a step that certainly has to be carefully considered. This appears to be an area of research that is mainly technology-driven. In their eminent paper on theoretical issues in today's man–machine systems research and applications, Johannsen, Levis *et al.* (1992) emphasize that much effort has gone into the design of all kinds of decision support systems, but that it is alarming to see that little time is spent on the evaluation and validation of these systems.

Even less is known about the application and usefulness of decision support systems in the petrochemical process industry than is the case in the area of nuclear power generation, the sector to which Johannsen, Levis *et al.* (1992) are referring.

The present paper is certainly not a plea to stop bothering about intelligent interfaces, but given the state of the art, it might be just as worthwhile:

1. to prevent alarm expansion resulting from superfluous alarm points and unjustified priority levels;
2. to lower the alarm load by correcting oscillating alarm points; and
3. to provide task-oriented alarm overview information that facilitates the operator's analysis and understanding of the alarm condition.

If our data are typical of large plants in the process industry, and we think they are, successful efforts with respect to the three topics just mentioned would reduce the operator's alarm load with about 40–60 per cent and would provide him with effective alarm analysis tools, without resorting to added complexity and costs of the process control system.

References

Alty, J.L. and Bergan, M., 1992, The design of multimedia interfaces for process control, in *Proceedings of the 5th IFAC/IFIP/IFORS/IEA Symposium on Analysis, Design and Evaluation of Man–Machine Systems*, the Hague, the Netherlands, June.

Bainbridge, L., 1987, Ironies of automation, in Rasmussen, J., Duncan, K. and Leplat, J. (Eds) *New Technology and Human Error*, Chichester: Wiley, pp. 271–83.

Dallimonti, R., 1972, Future operator consoles for improved decision making and safety, *Instrumentation Technology*, **19** (8), 23–8.

Dallimonti, R., 1973, New designs for process control consoles, *Instrumentation Technology*, **20** (11), 48–53.

Flippance, J.W. and Shaw, J.A., 1978, The operator-automation system interface, symposium: *The operator instrument interface*, Middlesbrough.

Fraser, G.L., 1978, A new approach to the interface between operator and the control system, symposium: *The operator instrument interface*, Middlesbrough.

Hal, G. van and Wendel, I.E.M., 1992, User-oriented design of man–machine interfaces; the design of man-machine interfaces for a processing line: a case study, in *Proceedings of the 5th IFAC/IFIP/IFORS/IEA Symposium on Analysis, Design and Evaluation of Man–Machine Systems*, the Hague, the Netherlands, June.

Hoonhout, H.C.M. and Zwaga, H.J.G., 1993, Operator behavior and supervisory control systems in the chemical process industry, in *Proceedings of the 5th International Conference on Human–Computer Interaction (HCI International '93)*, Orlando, USA, August, Amsterdam: Elsevier.

Johannsen, G., Levis, A.H. and Stassen, H.G., 1992, Theoretical problems in man–machine systems and their experimental validation, in *Proceedings of the 5th IFAC/IFIP/IFORS/IEA Symposium on Analysis, Design and Evaluation of Man–Machine Systems*, the Hague, the Netherlands, June.

Kortlandt, D. and Kragt, H., 1980, Process alerting systems as a monitoring tool for the operation, in *Loss Prevention and Safety Promotion in the Process Industry*, Vol. 3, pp. 804–14.

Marshall, E.C., 1992, Involving the user in the design of computer-based displays in power plant control rooms, in Lovesey, E.J. (Ed.) *Contemporary Ergonomics 1992, Proceedings of the Ergonomics Society's 1992 Annual Conference*, London: Taylor & Francis.

Rasmussen, J. and Vicente, K.J., 1989, Coping with human errors through system design: implications for ecological interface design, *International Journal of Man–Machine Studies*, **31**, 517–34.

Schaaf, T.W. van der, 1989, Redesigning and evaluating graphics for process control, in Salvendy, G. and Smith, M.J. (Eds) *Designing and Using Human–Computer Interfaces and Knowledge Based Systems*, pp. 263–70, Amsterdam: Elsevier.

Swaanenburg, H.A.C., Zwaga, H.J. and Duijnhouwer, F., 1988, The evaluation of VDU-based man–machine interfaces in process industry, in *Proceedings of the 3rd IFAC/IFIP/IFORS/IEA symposium on Analysis, Design and Evaluation of Man–Machine Systems*, Oulu, Finland, pp. 100–6.

Zwaga, H.J. and Veldkamp, M., 1984, Evaluation of integrated control and supervision in the process industries, *Institute of Chemical Engineers Symposium Series*, No. 90, 133–46.

Zwaga, H.J.G., 1993, The use of colour in CRT displays for the process industry, in Lovesey, E.J. (Ed.) *Contemporary Ergonomics 1993*, pp. 278–83, London: Taylor & Francis.

Part 3
Design and evaluation of
alarm systems

Design and evaluation of alarm systems

Neville Stanton

This section presents three chapters based on the design and evaluation of alarm systems. Chapter 8 (by David Usher) presents the alarm matrix: a representational form of the different aspects of system parameters. David illustrates how the matrix can be used to present required values, measured values, errors, tolerances and criticalities. He suggests that the matrix provides a means for rigorously defining alarm information. David argues that the provision of a definition of the alarm entity reduces the problem of alarm presentation to one of designing it in a manner so that it may be readily assimilated by the human operator.

This section presents three chapters based on the identification of the needs of the operator in alarm handling tasks. Chapter 9 (by Andreas Bye, Øivind Berg and Fridtjov Øwre) considers a variety of computer-based systems that assist the human operator, for example: alarm filtering, early fault detection and function-oriented plant surveillance. The authors propose that these methods complement each other in different plant operating regimes and provide diversity in plant monitoring systems. Andreas, Øivind and Fridtjov describe their experiences gained from the development of prototypes and actual plant installations. They also illustrate how a combination of alarm principles can be combined into an integrated alarm system. The integrated alarm system consists of three main blocks for:

- alarm generation;
- alarm structuring;
- alarm presentation.

The authors propose to build an alarm system toolbox to facilitate the implementation of specific alarm systems for different plants.

Chapter 10 (by Ned Hickling) describes the principles of design and operation of the alarm system to be used in the main control room at Sizewell

'B' pressurized water reactor nuclear power plant. Ned presents the ergonomics and engineering principles which are being used to achieve an operationally effective alarm management scheme. It is obviously desirable that the systems offer engineering redundancy and diversity given the safety critical nature of a nuclear power plant. Ned introduces the main ergonomics design principles important to the development of the control room, which are:

- careful consideration of operational needs;
- careful classification and prioritization of alarms;
- clearly assigning and coding alarm ownership;
- ensuring that visible alarms are relevant to the task;
- using hardware and software logical alarm reduction;
- the application of clear audible and visual coding;
- ensuring compatibility of the man–machine interface between different alarm states.

Ned outlines some current technical and operational advantages and limitations of high integrity applications (i.e. nuclear power) and presents his views on technological problems that need to be solved in order to achieve further improvements in the performance of alarm systems.

8

The alarm matrix

David M. Usher

Introduction

There is a lack of precision in the use of the term alarm both in the literature and in common parlance (Stanton and Booth, 1990). The term is sometimes used to describe part of a control desk (the alarm panel), sometimes to refer to a system state (the alarm is over), sometimes as a noun (cancel that alarm) and sometimes as a verb (that reading should be alarmed).

Many observational studies have revealed alarm handling to be the major part of the task of many plant operators. Indeed, the presentation of alarm signals is crucial to the successful operation of an industrial control room. Against this background it is surprising that no standard definition has been adopted by the human factors community.

It is the purpose of this paper to define an entity, to which the word 'alarm' can be applied, in such a way that our preconceptions of the meaning of the word are not violated. Once armed with a rigorous definition, the design of the alarm component of the human–machine interface should be simplified, not least because of the enhanced potential of computer algorithms.

The matrix model

In this model, each aspect of a time dependent system is represented as a matrix, with a column for each parameter characterizing the system, and a row for each stage of the evolution of the process. The rows of an aspect matrix represent therefore the state of that aspect of the system at each stage. In the case of continuous processes, the stages can be associated with the passage of time.

The aspect matrices

Within this formalism we may define for any process the 'required value'
matrix **R** of which the element r_{ij} specifies the value that system parameter
i is required to take at stage j of the process. Similarly, the element t_{ij} of
the 'tolerance' matrix **T** represents the maximum permissible deviation of
element ij from r_{ij}.

As the process unfolds, the matrix **M** of measured values is populated, and
so is the matrix **E** of errors, whose elements are given by the deviations of
the parameters from their required values, or:

$$e_{ij} = |m_{ij} - r_{ij}| \tag{8.1}$$

Let us now consider a matrix **A** whose elements are defined by the equation:

$$a_{ij} = B(e_{ij} - t_{ij}) \tag{8.2}$$

where $B(a) = 1$ if $a > 0$ and $B(a) = 0$ otherwise. The elements of **A** are set
to 0 when the system is close enough to its required state and 1 when it has
deviated too far.

Example – part 1

As an illustration of the method, let us consider the process of starting a
motor car and driving it away. The requirements matrix for this process
would contain a row representing the state of the car during each stage of the
activity, and a column for each feature of the car under scrutiny. For brevity,
in this example we will assign values only to seven rows and four columns of
the requirement matrix, as follows:

	Interior light (%)	Ignition lamp (%)	Speedometer (mph)	Motion
Approach car	0	0	0	0
Open door	100	0	0	0
Close door	0	0	0	0
Ignition on	0	100	0	0
Start engine	0	10	0	small
Move away	0	reduction	increase	increase
Steady speed	0	0	speed-related	steady

We observe that the elements of the requirements matrix need not be fixed
values, but may be functions of the other elements, such as 'increase'. Indeed,
this would be very likely in any non-trivial process. The corresponding tolerance matrix **T** might contain the elements:

	Interior light (%)	Ignition lamp (%)	Speedometer (mph)	Motion
Approach car	0	0	5	0
Open door	10	0	5	0
Close door	0	0	5	0
Ignition on	0	10	5	0
Start engine	0	10	5	0
Move away	0	10	5	small
Steady speed	0	10	10% of speed	small

As the process develops (in this case, the starting of the car), the measured values matrix **M** is populated. After the passage of the first five stages, it might contain the following elements:

	Interior light (%)	Ignition lamp (%)	Speedometer (mph)	Motion
Approach car	0	0	2	0
Open door	0	0	2	0
Close door	0	0	2	0
Ignition on	0	100	2	0
Start engine	0	100	2	0

and we may derive the corresponding elements of the error matrix **E** using equation 8.1:

	Interior light (%)	Ignition lamp (%)	Speedometer (mph)	Motion
Approach car	0	0	2	0
Open door	100	0	2	0
Close door	0	0	2	0
Ignition on	0	0	2	0
Start engine	0	90	2	small

The matrices **M** and **E** are not fully populated because the process did not proceed to completion. Using equation 8.2, the matrix **A** may be derived from **E** and **T**, as follows:

	Interior light	Ignition lamp	Speedometer (mph)	Motion
Approach car	0	0	0	0
Open door	1	0	0	0
Close door	0	0	0	0
Ignition on	0	0	0	0
Start engine	0	1	0	1

From this matrix we may observe that:

- the interior light did not function as it was supposed to when the door was opened;
- when starting the engine, the ignition warning lamp indicated a fault;

- insufficient motion was detected when the engine was started; and
- the non-zero speedometer reading was not considered important.

This is status information of the kind usually gleaned from an alarm system, and the contents of the matrix **A** can be seen to take on the character of alarms. Data generated in this way indicate when system parameters deviate unacceptably from the values they are required (or expected) to take at each stage of a process. Hence, the matrix **A** contains information only about unexpected events.

The alarm matrix

The treatment so far has yielded an alarm matrix whose elements are binary in character. But inherent in the broad concept of an alarm is a measure of its urgency or importance. Even the most unsophisticated alarm indicator systems will endeavour to add a further layer of discrimination by (for example) flashing a lamp to indicate a 'high-priority alarm'. In the example above, there is no indication that it was not the failure of the ignition lamp that caused the car journey to be abandoned rather than the lack of motion in the engine.

Therefore the input data must include an additional aspect matrix containing the importance of the system parameters to the prosecution of the process. We shall call this the criticality matrix **C,** whose elements c_{ij} represent the importance for the transition of the system from stage j to stage $j + 1$ that process parameter i should take its required value, r_{ij}.

These criticality data may be incorporated into the definition of the elements of the matrix **A** (equation 8.2) as follows:

$$a_{ij} = c_{ij} \cdot B(e_{ij} - t_{ij}) \tag{8.3}$$

Example – part 2

Bearing in mind that the process used as the subject of the example above is specifically that of starting and driving away a car, rather than carrying out a maintenance audit (as one automatically does), the criticality matrix might be of this form:

	Interior light	Ignition lamp	Speedometer	Motion
Approach car	0	0	0	0
Open door	0	0	0	0
Close door	0	0	0	0
Ignition on	0	100%	0	0
Start engine	0	0	0	100%
Move away	0	0	0	100%
Steady speed	0	30%	50%	100%

where the non-zero values at the 'steady speed' stage reflect respectively the importance of charging the battery, of knowing one's speed, and of continuing to move. The choice of granularity is of course arbitrary; a scale of 1 to 10 would be equally satisfactory.

The alarm matrix resulting from the application of equation 8.3 would be as follows:

	Interior light	Ignition lamp	Speedometer	Motion
Approach car	0	0	0	0
Open door	0	0	0	0
Close door	0	0	0	0
Ignition on	0	0	0	0
Start engine	0	0	0	100%

From this matrix we can see that the effect of the inclusion of the criticality matrix **C** has been to prioritize the alarms, and to illuminate the nature of the fault, since it is now clear that there is nothing amiss with the ignition system: the fault must lie with the starter motor itself.

It is the contention of this chapter that the contents of the aspect matrix **A** defined by equation 8.3 can be considered to be the alarms generated by a process.

Retrospection

The plant operators' task is of course to use their skill, training and knowledge of the process to attempt to determine the system state from the data presented to them in the control room. Clearly the alarm data form a small part of the total information at their disposal, and provide only a coarse guide in the identification of a fault (Rasmussen, 1974). However, it is clear that in many cases it is the onset of an unexpected event, signalled by an alarm annunciator, that initiates the diagnostic task. To discover the plant state from the alarm data is to travel in the reverse direction from that taken in the above analysis and attempt to derive the matrix **M** from the matrix **A**.

In many industrial control-rooms the various aspect matrices for the process are built into the process control machinery, for example in a computer program, but in other cases they reside wholly in the operator's mind.

Realisation

It is the task of the human factors specialist to realize the alarm matrix for a particular process in such a way that the operator can extract the information needed to regress through the various aspect matrices and diagnose the problems that might have occurred on the plant.

It is obvious from this treatment that very considerable process knowledge is contained in the aspect matrices **R**, **T** and **C** and that this must be elicited before any attempt can be made to derive the alarm matrix **A**. The implementation of the alarm display panel will depend to a large extent upon the results of this derivation, depending as it does on the number of non-zero elements in **A**. Clearly the alarm aspect of a system is intrinsic to it and cannot be grafted on as an afterthought, whether the display of alarm data is computerized or mechanical.

Once an expression of the alarm aspect has emerged, however, the task of realization need not be onerous, since although the aspect matrices themselves may be very large in a complex system, the number of their non-zero elements is likely to be at least an order of magnitude smaller than the number of indicators on a traditional alarm panel.

In this context it is worth reiterating that the term 'alarm' does not describe pieces of equipment. Alarms are data, in the same class as temperatures and pressures. Therefore the word alarm should *not* be used to refer to an indicator, an annunciator, or any other hardware used to communicate with plant operators, by analogy with the importance of preserving the distinction between a temperature and a thermometer.

Any practical application of this method will have implications for the system instrumentation. It has been seen that in order to derive alarm data, it is necessary to establish which stage in the process has been reached. In many process control situations this may be difficult to gauge and require the investment of considerable effort in the development of computer algorithms. A simpler solution would be, on the occasions when the automatic determination of the process stage has failed, to exploit the knowledge and training of the operators and allow them to enter their perception of it into the alarm display system.

Another problem is posed by the entry of the tolerance data, which because they depend on the process stage, will be much larger in number than is the case in traditional alarm systems. Moreover, it is unlikely that they can be specified with sufficient accuracy when the plant is being designed. Here, the solution must be to allow users continuously to amend the tolerances in the light of operational experience. Systems might be envisaged which automatically adjust the tolerance data on the basis of the choices that operators have made of alarms to be displayed or shelved.

Finally, it should be recalled that the criticality matrix **C** must also be populated, and that the choice of criticality matrix is determined by the current task. A maintenance engineer does not require the same output from the alarm system as the shift-worker. For this reason, the human–machine interface of the alarm system should allow the users to specify the nature of the current task.

Conclusion

The alarm matrix has been derived from an analysis of the time dependence of a system. It is seen to depend upon matrices defining the required values, their tolerances and their importance to the process. The analysis yields a rigorous definition of the term 'alarm' to assist in the design of the human–machine interface in an industrial context.

References

Stanton, N.A. and Booth, R.T., 1990, The Psychology of Alarms, in Lovesey, E.J. (Ed.) *Contemporary Ergonomics*, pp. 378–83, London: Taylor & Francis.
Rasmussen, J., 1974, Communication between operators and instrumentation, in Edwards, E. and Lees, F.B. (Eds) *The Human operator in Process Control*, London: Taylor & Francis.

9

Operator support systems for status identification and alarm processing at the OECD Halden Reactor Project – experiences and perspective for future development

Andreas Bye, Øivind Berg, Fridtjov Øwre

Introduction

The OECD Halden Reactor Project has for several years been working with computer-based systems for determination of plant status including alarm filtering, early fault detection, and function-oriented plant surveillance. The methods explored complement each other in different plant operating regimes and provide diversity in plant monitoring systems. The work has been carried out by development of prototypes in the HAlden Man–Machine LABoratory HAMMLAB and in installations at nuclear power plants.

This paper describes the characteristics of the various methods explored at the Project and the experience gained from actual plant installations. A combination of different alarm principles into a new integrated alarm system is discussed. The integrated alarm system consists of three main functional blocks for 1) alarm generation, 2) alarm structuring and 3) alarm presentation. It is proposed to build an alarm system toolbox to facilitate the implementation of specific alarm systems for different plants.

One of the main tasks for operators in nuclear power plants is to identify the status of the process when unexpected or unplanned situations occur. The alarm system is the main information source to detect disturbances in the process, and alarm handling has received much attention after the Three

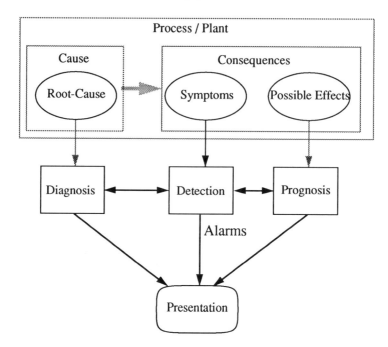

Figure 9.1   Detection, diagnosis and prognosis systems versus an abstract view of the process. The thick, shaded arrow indicates an imagined cause–consequence sequence, the full-drawn arrows indicate data flow, while the thin, dotted arrows indicate that the diagnosis and prognosis modules try to find the root-cause and the possible effects, respectively.

Mile Island accident (Kemeny, 1979). Here it was realized that conventional alarm systems created cognitive overload for the operators during heavy transients.

Disturbance detection systems in the form of alarm systems are present in a wide variety today, and the OECD Halden Reactor Project has been working with different methods in several systems. Filtering and handling of conventional alarms were treated in the HALO (Handling Alarms using LOgic) system (Øwre and Marshall, 1986). Model-based alarm methods were explored through a system which tries to detect disturbances at the earliest possible stage, EFD (Early Fault Detection). (Sørenssen, 1990). A function-oriented approach was first explored through CFMS (Critical Function Monitoring System) and SPMS (Success Path Monitoring System), and later through the post trip guidance system SAS-II (Øwre, Nilsen *et al.*, 1991).

In addition to alarm handling and monitoring systems other methods have emerged during the last 10 years to support the operator in his/her status identification task. Examples are diagnosis and even prognosis systems. In Figure 9.1 an abstract representation of a process in disturbance is presented, with different surveillance systems for detection, diagnosis and prognosis.

Basically, there are only causes and consequences. Consequences are by definition occurring after the causes.

- 'Root-cause' is the primary cause of the disturbance in the plant. It may or may not be directly detectable (observable) through the available process instrumentation. Often, the root-cause will only be detectable through its consequences.
- 'Symptoms' constitute the set of consequences of the root-cause which at a given time are directly detectable through the process instrumentation.
- 'Possible effects' constitute the rest of the set of consequences of the root-cause, i.e. not at the moment detectable consequences, and future/potential consequences. Automatic or manual actions must be taken to prevent dangerous effects.

In the detection block of Figure 9.1, the symptoms of the root-cause will always show up, sooner or later, by definition. If no mitigating control actions, either manual or automatic, are taken in time, at least some of the possible effects will appear and may eventually be detected. The diagnosis block tries to diagnose the root-cause of the disturbance, while the prognosis block tries to predict possible effects. Both of these systems use information from the detection part, as well as other process data. The prognosis block could use data from the diagnosis block also. Suggested root-causes from the diagnosis block and suggested possible effects from the prognosis block are then presented to the operator for further interpretation.

If proper corrective actions are implemented in time, the future/potential consequences will never appear, and the root-causes and symptoms will disappear after some time, at least after repair or maintenance have been completed.

An alarm system normally resides within the detection block in Figure 9.1. However, it can be very difficult to separate the three blocks, because the methods for detection, diagnosis and prognosis may overlap. Anyhow, the above structure may serve as a clarifying picture for developers.

Experience with existing systems at the project

Handling of conventional alarms

Alarm handling is an area which has received much attention and the OECD Halden Reactor Project has developed an alarm filtering system called HALO using logic filtering to reduce the number of active alarms during process transients. Filtered alarm information is presented by means of colour cathode ray tubes (CRTs) in a hierarchical display structure. In order to make a thorough evaluation of the HALO system, it was implemented on the Project's full-scope Pressurized Water Reactor, Nokia Research Simulator (PWR NORS) which has a realistic alarm system with a potential list of around 2500

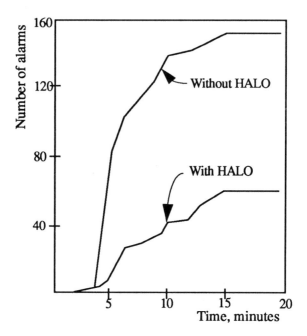

Figure 9.2   The number of alarms issued in a transient with a small leakage in the primary circuit. Note that the degree of filtering at the beginning of this transient is rather low.

messages. Two series of experiments have been carried out, and the main results are described below.

The first experiment concentrated on two issues (Marshall and Øwre, 1986): whether a reduction in the number of alarms presented to the operator helps him/her to interpret and handle a process transient, and how best to present alarm information utilizing the flexibility of a computer-based system. The first issue was tested by comparing the response of operators when presenting a filtered text alarm list with an unfiltered text alarm list. The second issue was tested through a two-level alarm interface in addition to the process formats. The degree of filtering for one of the transients chosen is illustrated in Figure 9.2.

A significant difference between the amount of unfiltered and HALO-filtered alarms is present. However, the experimental results showed few differences in operator performance between presentation of the filtered versus the unfiltered alarm lists. The scenario was criticized as being a too 'kind' transient, where there was not too much difference between the filtered alarm list and the unfiltered, especially in the initial phase. This phase proved to be the most important for the operators' investigation process, and it was observed that the operators did not always attend to new alarms. To improve this it was argued that an alarm display should 'insist' that the operator attends to new alarm conditions.

The man-machine interface (MMI) was criticized as being slow and cumbersome, because the operators had to path through a group alarm level between the overview picture and the process formats. The subjects also wanted more process information included in the overview picture.

The second HALO experiment explored an improved NORS/HALO system, based on the findings of the first experiment, and compared this system with a conventional system (Marshall, Reiersen *et al.*, 1987). The improved HALO system excluded the intermediate group alarm formats on the level between the overview and the process formats. The alarm information on this group alarm format was either included in the overview, in the process formats or completely removed.

Alarm information was thus embedded in the process formats. Process information was also included in the overview picture, as proposed by the operators in the first experiment. A third change was that a blinking alarm on the overview could only be accepted by calling up the appropriate process format, thereby forcing the operator to a larger extent to take notice of new alarms.

The new alarm presentation with only two levels of information, the overview and the process formats, was regarded as much easier to use than the three level approach. The alarm information on the process formats helped the operators to locate disturbed plant variables. The subjects preferred the new alarm system, and found it easier to use compared to the conventional system, especially when a large number of alarms were active.

Model-based alarms

Fault detection based on static alarm limits is difficult to apply to dynamic processes. In order to get an early alarm warning and thus avoid the taking of drastic countermeasures to restore normal plant conditions, the alarm limits should be put very close to the desired operating points. However, this is difficult in practice for a dynamic process, because a certain operating range for the process variables is usually required.

Another difficulty is the fault propagation in a complex, closely-coupled process. The fault may have its origin far away from the point where the first alarm is triggered. This leads to difficulties when the operator tries to find the cause of the alarm, since many systems may be disturbed. Shortly after the first alarm has triggered, a number of other alarms will usually trigger, too. It may be difficult for the operator to tell whether these alarms mean new faults or the same fault once more. An approach to the diagnosis of faults based on such alarms has previously been outlined by the Halden Project (Bjørlo, Berg *et al.*, 1987).

As an extension of several years' activity on alarm reduction methods the OECD Halden Reactor Project started in 1985 to develop an early fault detection system. The method used is to run small, decoupled mathematical

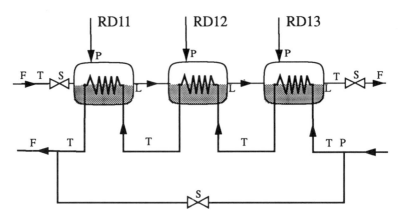

Figure 9.3   A preheater train with available measurements.
P = pressure, T = temperature, F = flow, L = level, S = valve position.

models which calculate the state of the process assuming no faults in parallel with the process. The behaviour of these models are then compared with the behaviour of the real process, and if there is a deviation, an alarm is triggered. In this way derived alarms are avoided, and one will only get one alarm for one fault.

Prototypes of the EFD system developed for simulators and installations in real power plants e.g. the Imatran Voima-owned plant Loviisa in Finland have demonstrated the feasibility of this methodology (Sørenssen, 1990). An example is given here for the high pressure preheaters.

High-pressure preheaters in many power plants are subject to corrosion and leakages. Such leakages reduce the thermal efficiency of the plant. They are difficult to detect, as water only flows from one place in a closed loop to another place in the same loop. Figure 9.3 shows the available measurements of a preheater train. One would have liked to have a few more, but as measurements are expensive to install in an existing plant, it was decided to use only what was available. For each preheater three balances were established: a mass balance for the tube side, a mass balance for the shell side and a single heat balance for both sides.

The flow of water into the shell side is measured only in the preheater denoted RD11. Using the RD11 measurements and the balance equations, the flow out of the shell side of RD11 can be calculated. Combine this calculation for the flow of water into the shell side of RD12 with RD12 measurements and the exercise can be repeated for RD12, and the flow found from RD12 to RD13. Another repetition, and we have the flow out of RD13. Because the flow out of RD13 is measured as well as calculated, the two can be compared. If the deviation exceeds a certain limit, an EFD alarm is issued. There are three levels of alarms, depending on the size of the deviation. Except for the alarm limits, the models contain no adjustable parameters.

Figure 9.4, taken from the RD10 train in Loviisa unit 1, shows a situation

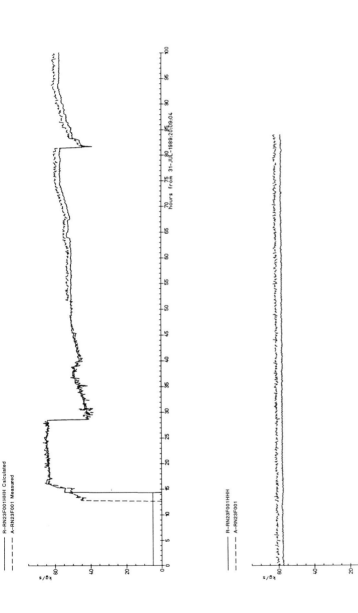

Figure 9.4 The initial agreement survives the jump at 29 hours. A small deviation appears at 52 hours, increasing with time. The deviation survives the jump at 82 hours. Detection of leak.

with dynamics, the start-up after the summer stoppage. At about 14 hours the preheaters are switched on. (The horizontal tracing indicates that the preheaters are bypassed and not in operation). At 52 hours a small deviation shows up and increases slowly up to the end of the series at 184 hours. We notice that the agreement between the calculated and the measured curves survives the large jump at 29 hours, while the disagreement survives the large jump at 82 hours. The reason for this deviation is a small leak.

It would have been very difficult for an operator to discover the small deviation which occurred at 52 hours in Figure 5.4 just by observation. None of the single measured variables make a big jump, it is only when all are considered together that the phenomenon shows up. The big jumps of almost all variables at 29 hours and 82 hours, however, do not indicate a change in the preheaters themselves, only in their environment.

The fact that the models describe the plant through dynamic as well as static situations, and with good accuracy, means that it is possible to use rather narrow alarm limits. This means that we have a sensitive system which can detect faults while they are still small.

Function-oriented alarms

In case of major disturbances in a plant with a large number of alarms issued, a function-oriented approach is used in many cases to monitor plant status. Instead of looking at single systems or variables and alarms within a system one monitors critical safety functions in terms of whether these functions are challenged in a major disturbance. The OECD Halden Reactor Project investigated the critical safety function concept in several studies in the period from 1983 to 1987 in co-operative projects with Combustion engineering in the US and the Finnish utility Imatran Voima. In particular, the human factors experiment with the Success Path Monitoring System (SPMS) clearly showed distinct improvements in operator performance with respect to appropriate corrective actions being taken in disturbance situations (Baker, Marshall et al., 1988).

The OECD Halden Reactor Project has participated in development of a function-oriented advisory system called SAS-II to assist the shift supervisor in his/her observation and evaluation task after plant disturbances leading to emergency shutdown. To monitor this emergency shut-down process the supervisor today applies a set of function-oriented emergency procedures. SAS-II will give continuous information to support the work of the emergency procedures. As well as giving alarm it will explain why critical safety functions are challenged (Øwre, Nilsen et al., 1991). SAS-II is a joint research programme between the Swedish nuclear plant Forsmark, the Swedish Nuclear Inspectorate and the OECD Halden Reactor Project.

The SAS system is a very high level function-oriented alarm system. There are a total of 12 different alarms in the present version of SAS. These 12 alarms describe the status of four critical safety functions. There is a wide

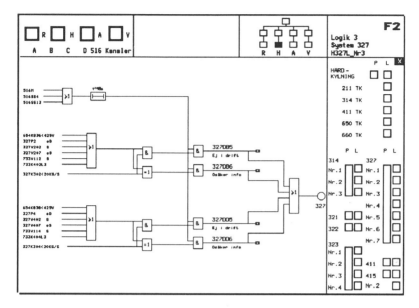

Figure 9.5   Example of a logic diagram in the user interface of the SAS-II system.

variety of possible reasons for one critical function to be challenged, and the alarms are defined in terms of very complex logic diagrams. These logic diagrams are not only the definition of the alarms. They are also a part of the interface to the operator, as shown in Figure 9.5.

The large field shows the logic diagram, and by using colours to reflect false/true status of the logic gates the operator will be able to follow a red path back to the causes of an alarm. The user interface also illustrates where this logic diagram is situated in a diagram hierarchy, and it shows the overview alarms in the upper left corner.

When SAS-II is finally taken into operation action it should be an improvement to safety for a number of reasons: on the computer screen at his/ her own work place, the shift supervisor will easily and clearly get the information he/she needs when applying the emergency operating procedure. Also the computerized system will warn the supervisor if any of the defined critical safety functions are challenged, both during normal operation and in particular after emergency shutdown, and explain why.

A new integrated alarm system

Alarm processing

The various systems described in the previous sections provide the basis for development of a new integrated alarm system, where the alarm processing is considered to be a three-stage process, as given in Figure 9.6.

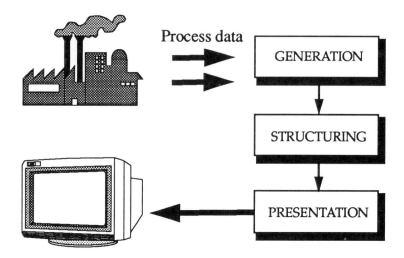

Figure 9.6 Alarm processing in three stages: alarm generation, alarm structuring and alarm presentation.

The first stage takes care of the generation of the whole spectrum of alarms. These do not need to be only component- or measurement-related alarms, but can equally well be higher level functional alarms like 'core heat removal challenged'.

The structuring stage includes alarm filtering, where a number of filtering methods are available. However, to support the operator in the status identification task, types of structured alarm lists other than those used for standard filtering should be available. The operator should be able to access exactly those alarms he needs to see for support in the diagnosis task. This structuring should be far more sophisticated than the after/prior relationship and the more/less important relationships used up to now. This kind of structuring will normally not divide the alarms into digestible chunks of information.

The last stage is the presentation stage. It is related to the structures established in the previous stage and it should seek to present those structures in an optimal way. Both permanent and selectable structures/alarm lists should be available. It should be easy to switch from one alarm structure to another in selectable displays.

Alarm generation

Several types of alarms should be present in a new alarm system. Alarm generation is the phase where all new alarms are generated from process measurements. This includes all conventional alarms present in a plant, in addition to the new types of alarms sketched in the preceding sections, model-based alarms as in EFD and function-oriented alarms as in SAS-II.

COMPLETE ALARM-LIST

FILTERED ALARM-LIST

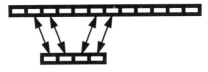

Figure 9.7   The filtered alarm list is generated from the complete alarm list. The arrows pinpoint the common alarms.

The generation of conventional alarms will require a limit check module only. In addition to the normal static approach, dynamic limits may be applied as well. This alarm generation function is often done in the process control systems and, therfore, it must be possible to connect a new alarm system to already existing alarm or process control systems. Then parts of the alarm generation will be outside the new system, and only structuring, e.g. filtering, and presentation, will be taken care of by the new alarm system.

Alarm structuring

One purpose of alarm structuring is to keep the number of presented alarms on a level where one can avoid cognitive overload of the operators, and create cleaner alarm pictures without false alarms. Alarm structuring has to do with alarm filtering and presentation, i.e. the amount of alarms presented whenever a disturbance or a dynamic process transient occurs. The HALO system, described above, is an example of an alarm system addressing this point. Another purpose with alarm structuring is to help the operator in his/ her task to diagnose disturbances in the process. This may be done with a flexible alarm system, which the operator can use interactively, choose different alarm structures, and thereby get a better overview of the situation.

Filtering

Filtering can be seen as a special case of structuring. Loosely defined, alarm filtering can be seen as removing the unimportant alarms from the complete set of alarms. Graphically, this can be seen as the structure shown in Figure 9.7. The arrows indicate which alarms are the same. The operator normally sees the filtered alarm list and the complete alarm list will only be printed out on some kind of event log.

Some types of conventional alarms do not provide relevant information or do not contain information about new disturbance situations, so they should be filtered. Many methods can be used to support this. HALO introduces and implements a number of such methods. (Øwre and Tamayama, 1982.)

Repetitive alarms caused by signal noise do not contain information about new disturbances. *Lowpass filtering* of analogue measurements may help to remove these false alarms. False alarms from binary signals, due to oscillations near the alarm limits, will be truncated effectively by introducing time delays or deadbands.

Precursor alarms should be filtered. When a HiLevel alarm is on, and a HiHiLevel alarm is triggered, too, then HiLevel is called a precursor to the latter. *Simple logical relations* may be used to filter these precursors.

Operational mode dependent 'false' (standing) alarms should be filtered. Some alarms may be relevant in full power production while not at all relevant in shutdown state. They are called standing alarms and are false ones because they do not inform about what is wrong in the process. Whenever alarms are gathered in *groups* of some sort, they may be filtered easily by relating them to discrete operational modes. The most common way of identifying the process modes is to check the status of some few important process parameter, e.g. power, flow rates or pressures. Relating the different alarms to the operational modes requires highly experienced plant engineers, and it is regarded as a time consuming job.

Consequence alarms may be defined as secondary alarms which are consequences of a disturbance indicated by another present alarm. In case of a disturbance in the process, a lot of conventional alarms are often triggered because of the propagation of the transient which is induced. Many of these offer irrelevant information or noise to the operator and they should be filtered or suppressed. Important consequence alarms should however not be filtered. Several related methods may be used for this purpose, e.g. *first out*, where alarms are gathered in groups. The first triggered alarm is alerted in the ordinary way, while the rest of the alarms in the group are suppressed. *Structured hierarchy* is a method where alarms are gathered in groups which internally are structured in a prioritized hierarchy. However, these methods may also be used for structuring into other types of lists, which the operator can use interactively in his/her diagnosis task.

Flexible structuring

One method which could be used for such structuring is *causal trees*, where alarms are organized in 'trees', which give causal relations between the alarm tags. This means that some alarms are viewed upon as 'cause alarms', and some are looked at as 'consequence alarms'. Alarm-lists of all alarms which are causal predecessors of some particular alarm is shown in Figure 9.8.

The idea is then to give the operator a MMI so that he/she can move around in this structure by following the arrows. Looking at the filtered alarm list it should be possible (by means of one mouse click) to enter the predecessor list for one of the alarms. It should also be possible to follow the double arrows, i.e. moving from the occurrence of an alarm in one list to the occurrence of the same alarm in another list. In order to construct such lists it will be necessary to have quite complex information about the interdependencies of alarms. However, it is not necessary to require that these interdependencies are 100 per cent accurate. It is believed that they will be useful even in the case where they are normative.

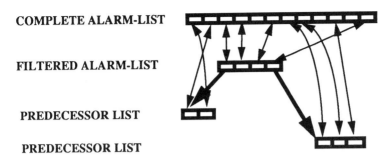

COMPLETE ALARM-LIST

FILTERED ALARM-LIST

PREDECESSOR LIST

PREDECESSOR LIST

Figure 9.8 Structuring alarms into different types of alarm lists. The double arrows indicate identical alarms. The directed arrows indicate causal predecessors.

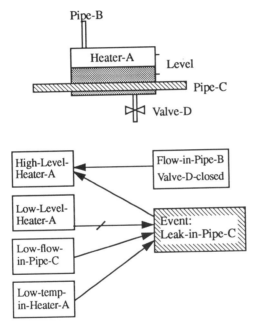

Figure 9.9 Event/alarm network for hypothesis related alarms. In this network the shaded box represents a non-observable event, and the others represent alarms or combinations of alarms.

An extension of this concept could be to include non-observable events in the causal tree. This makes the alarm system useful in fault diagnosis. Let us exemplify this by considering an example given in Figure 9.9.

The arrows without the crossing line indicate preceding/succeeding alarms or events, while the arrow with the crossing line indicates the opposite, i.e. that a *low-level-heater-A* alarm probably did not precede the event *leak-in-pipe-C*.

A list of predecessor-pairs is specified for the alarm *high-level-heater-A* in the example in Figure 9.9:

high-level-heater-A:

p-p: ((flow-in-pipe-B, valve-D-closed),
        (leak-in-pipe-C, void))

Flow-in-pipe-B combined with a *valve-D-closed* 'alarm' may then explain why the alarm *high-level-heater-A* has appeared. *Leak-in-pipe-C* refers to a non-observable condition. This hypothesis could be the diagnosis if no other 'alarms' are present.

The operator can make a hypothesis, this can either be an alarm which has not yet occurred, or a non-observable event. The interdependency relationships can be used to confirm or refute this hypothesis. The hypothesis must be present in the hierarchy either as an unobserved alarm or as a non-observable event. For example, alarm lists could be set up for the hypothesis *leak-in-pipe-C*:

leak-in-pipe-C:

Confirming events:
12:03:15 Low-flow-in-pipe-C

unconfirming events:
11:45:20 Low-level-heater-A

Not yet observed events:
??:??:?? Low-temperature-in-heater-A

Of course, since the relations are only normative, the presence of both confirming and unconfirming alarms cannot be excluded. For instance if both *low-flow-in-pipe-C* and *low-level-heater-A* have occurred, the operator may ask himself whether really *leak-in-pipe-C* is true.

In this very simple example, it is straightforward to overview the relations. But in a realistic situation, the number of relationships and active alarms will be too many to overview. It would be useful if the system could offer the operator some help in selecting those alarms which seem to confirm or refute his/her hypothesis. In this situation the operator will use the network interactively on-line, by interrogating the system with respect to some particular hypothesis. This network could be subjected to a systematic search to find those hypotheses which have the best support from the active set of alarms.

Alarm presentation

The HALO experiments as described in former sections provided valuable experience both regarding alarm presentation and operator handling of alarms. One conclusion was that two levels of displays in a hierarchy was optimal. If three levels are used, the operators may find it too cumbersome to use all of them in stressed situations.

An overview of an alarm presentation module is shown in Figure 9.10. Three different types of displays are envisaged in the alarm presentation. At the top level a permanent display, which consists partly of mimic diagrams, presents overview information in the form of key alarms and key process variables. The notation 'key alarm' means that the alarm in question is judged by process experts to be very important for the operator as it gives a good overview of the process state at any time. In principle a key alarm can be

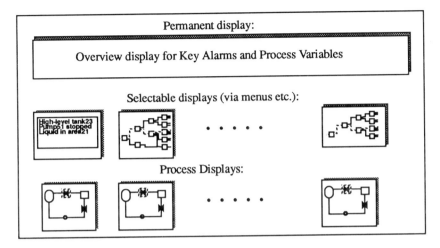

Figure 9.10 Alarm presentation. Permanent displays show important overview information. Selectable alarm displays are available to support the operator in his diagnosis task, while the process displays show both process and alarm information.

any kind of alarm: component alarm, group/system alarm, combined alarm, function alarm, or safety function alarm. The key process variables are defined in a similar way.

Preferably the overview status of key alarms and process variables should be shown on one large display, which must be easily readable from all the operator's working positions in the control room. If a large display is not available, a set of standard displays can be positioned close together in a group, thereby making it possible to form approximately the image of a corresponding large display.

In the overview display(s), the various key alarms could be sorted with respect to severity, type or function. Preferably the number of key alarms which may be present during disturbance and accident situations should never exceed a predefined limit. This should be tested by running anticipated transients on a full-scale simulator. The key alarm symbol must show, if there are several active alarms in the overview picture, which alarm that was activated first. This may be done by different darkness/brightness of the colours of the symbols. The key alarm symbols must be deactivated automatically, when the alarm situation returns to normal.

The selectable displays and the process formats are at the same level, so the process formats can be reached directly from the overview. The selectable displays should have a more independent role, and be available on a given number of screens. The displays are selected by the operator via menus etc. In principle, these displays could be constructed in any way depending on the end-user's demands. To be able to do that, the clue is in the alarm structuring. All the different structured alarm lists outlined in the structuring section may be selected. Examples of different structured alarm lists are: the complete alarm list, filtered alarm list, hypothesis alarm list, unexpected event list and a consequence supervision list.

As the HALO experiments indicate, alarms should be included in the process displays. Different symbols may have to be included to identify the different alarm types. Alternatively or in addition, the small alarm symbols can be made addressable, and by use of window-techniques, more information about the alarm can be shown directly in the process displays.

An alarm system toolbox

A general alarm system, which can be used on several different processes, should be flexible and expandible. Flexible meaning that the alarm system designer is able to make a specific alarm system out of the 'bits and pieces' provided by the general alarm system. Expandible meaning that the alarm system designer is able to expand the functionality of the general alarm system in order to support special functionality needed in a specific alarm system. Thus an important part of a general alarm system would be an alarm system toolbox, containing elements with the basic functionality needed to build specific systems.

Basic alarm elements are general, functional elements which include all the generating, structuring and filtering elements needed. Examples of such elements are: alarm grouping and prioritizing elements, model-based alarming elements, functional alarming elements, limit check elements, etc. If the basic alarm elements provided are not sufficient, the alarm system designer should be able to define his own alarm elements. In order to make a specific alarm system match a process, the designer must be able to put instances of the basic alarm elements together to suit the structure of the process, and to connect the alarm elements to process data.

An alarm system toolbox should contain the following:

- basic alarm elements;
- user defined alarm elements;
- methods for combining alarm elements;
- methods for making new user defined alarm elements;
- methods for connecting alarm elements to process data or to any other alarm element;
- methods for connecting external generated data to alarm elements.

In Figure 9.11 the small boxes (limit checks) are alarm elements supporting limit checks, where the designer can specify the alarm limits for connected process data. The large box is also an alarm element which contains user specified criteria for how to make a group alarm, based on individual alarms. In a similar manner a complete alarm system can be built and connected to process data.

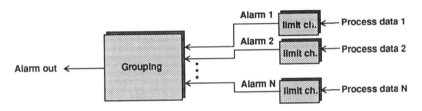

Figure 9.11   Alarm elements. A simple example of how a grouping alarm element and limit check alarm elements can be put together to produce one single group-alarm out of process data.

Conclusion

The three systems described, HALO, EFD and SAS-II, provide good methods and substantial background for developing a new integrated alarm system. HALO reduces the amount of presented alarms to the operator by a significant factor. The experiments with the presentation displays concluded that a two-level display hierarchy was preferred to a three-level approach. EFD utilizes models of the process in the detection of disturbances, and the implementation at the Loviisa nuclear power plant in Finland has proved that the method works. In the given example EFD is able to find internal leakages in heat-exchangers far earlier than other systems. SAS-II is surveying four critical safety functions at the Forsmark nuclear power plant in Sweden. It supports the shift supervisor's work with the emergency procedures, and warns him if critical safety functions are challenged, and explains why. Experiences from the installation are however not yet available.

In the new integrated alarm system, the alarm processing is divided in three main stages: alarm generation, alarm structuring and alarm presentation. A new alarm system must be able to generate all types of alarms, but it should also be able to import already generated alarms from other systems. The main difficulty with existing alarm systems is the cognitive overload of the operators in heavy transients. The alarm structuring tries to improve this problem by filtering the alarms presented to the operator. However, the structuring is also meant to support the operator with different types of alarm structures which he/she can use interactively in the status identification task in case of disturbances. As an extension into diagnosis support, non-observable events such as internal leakages may be added to the alarm structures. Three types of alarm presentation display are proposed: overview displays, selectable displays which the operator can use interactively, and the ordinary process displays.

The Halden Reactor Project explores a new integrated alarm system, i.e. a general system which includes the functionalities of the current alarm systems. To make such a new alarm system flexible and to make it possible to couple it to different processes, a proposal for an alarm system toolbox was made.

References

Baker, S., Marshall, E., Reiersen, C., Smith, L. and Gaudio Jr., P.J., 1988, The experimental evaluation of the Success Path Monitoring System – results and conclusions, *Halden Work Report, HWR-224,* May.

Bjørlo, T.J., Berg, Ø., Grini, R.E. and Yokobayashi, M., 1987, Early detection and diagnosis of disturbances in nuclear power plants, *Proceedings of the ANS Topical Meeting on Artificial Intelligence and Other Innovative Computer Applications in the Nuclear Industry,* Snowbird, Utah, USA.

Kemeny, J.G., 1979, 'Final report of the President's commission on the accident at Three Mile Island,' Washington, DC, USA.

Marshall, E. and Øwre, F., 1986, The experimental evaluation of an advanced alarm system, *Proceedings of the International ANS/ENS Topical Meeting on Advances in Human Factors in Nuclear Power Systems,* Knoxville, Tennessee, USA.

Marshall, E., Reiersen, C. and Øwre, F., 1987, Operator performance with the HALO II advanced alarm system for nuclear power plants – A Comparative study, *Proceedings of the ANS Topical Meeting on Artificial Intelligence and Other Innovative Computer Applications in the Nuclear Industry,* Snowbird, Utah, USA.

Øwre, F. and Tamayama, K., 1982, Methodologies for developing alarm logic in a HALO system, *Halden Work Report, HWR-80.*

Øwre, F. and Marshall, E., 1986, HALO – Handling of Alarms using LOgic: background, status and future plans, *Proceedings of the International ANS/ENS Topical Meeting on Advances in Human Factors in Nuclear Power Systems,* Knoxville, Tennessee, USA.

Øwre, F., Nilsen, S., Forsman, T. and Stenmark, J.E., 1991, An operator support system for a Swedish Nuclear power plant control room, *Proceedings, EPRI conference on Expert System Applications for the Electric Power Industry,* Boston, Massachusetts, USA.

Sørenssen, A., 1990, Early fault detection at the Loviisa Nuclear Power Plant by simulation methods, *Modelling and Simulation, Proceedings of the 1990 European Simulation Multiconference,* Nuremberg, Germany.

10

Ergonomics and engineering aspects of designing an alarm system for a modern nuclear power plant

E.M. Hickling

Introduction

The Sizewell 'B' nuclear power plant is the first in the UK using pressurized water reactor technology. The main control room is the centre for the control of plant operation, and will be manned by an operator and a supervisor. A second operator can be called in at busy times and during an incident or fault on the plant. The control room has within it three different interfaces which generate alarms concerning the plant. These are the alarm facia system, the discrepancy scheme and the distributed computer system (DCS).

This chapter seeks to illustrate that the careful application of ergonomics principles to alarm systems design, coupled with good procedural methods, can enhance the performance of alarm interface users.

The alarm facia system has an interface consisting of trans-illuminated labelled tiles. These alarms relate to around 35 critical safety parameters but around 60 tiles exist. Some information is duplicated or triplicated to provide signal redundancy. This ensures continued access to alarm information upon failure of independent data channels – known as separation groups. The tiles are arranged in a functionally-based layout which reflects the ordering of the control/display interface on the consoles.

The discrepancy scheme provides the control console switches used to start and stop equipment and realign circuits within the plant. Their orientation within mimic line diagrams of the plant indicates the state of the controlled item. Each switch uses an indicator lamp code to indicate any discrepancy between the switch orientation and the plant.

The network-driven distributed computer system is the principal interface for monitoring the state of the plant. It contains all the alarm information of the alarm facia system, and the plant equipment and process alignment information, including that represented by the discrepancy scheme. A comprehensive suite of task-related and system-related display formats is provided. These are disposed in a shallow and wide hierarchy consisting of mimic diagrams with embedded alarms, trends, bargraphs and alarm lists. There are also three permanently displayed plant overview formats. They are accessed by means of keyboard controlled VDUs distributed throughout the main control room.

Safety engineering principles

The level of safety required for Sizewell 'B' demands the use of information systems with high reliability and functional integrity. Three distinct alarm systems meet the requirements for information system redundancy and design diversity, to ensure that safe monitoring of the plant remains possible following an information system failure. Extensive engineering analyses of faults affecting safety have been performed to establish which process parameters will unambiguously or most sensitively enable the monitoring of the onset and progression of a fault.

For site licensing purposes, the reliability of the Alarm facia system is required to be the highest that can be reasonably achieved. This system annunciates only the most important safety related alarms which link to critical safety functions (CSFs) that have exceeded acceptable thresholds. The scheme of CSFs follows the general functional requirements laid down by the US Nuclear Regulatory Commission US NRC (1981), which resulted from the accident at Three Mile Island. A detailed description of them is given by Corcoran et al. (1981). The CSFs may be thought of as analogous to the vital life signs of the medical patient. They indicate the general condition of the patient but may not by themselves diagnose the disease. Like the vital life signs, symptomatic treatment of abnormalities in the CSFs is implemented in the absence of a fault diagnosis. The CSF displays are then used to monitor conditions in parallel with fault mitigation in the presence of a diagnosis.

To achieve very high reliability, invariant software (i.e. with no conditional logic) will be used. This reduces the possibilities for software failure but can give irrelevant alarms for some parameters in certain operational modes. The way this is handled by the user in operation is further described below.

The discrepancy scheme is also required to be highly reliable. Here simple software-based logic is used to detect the discrepancies between the indicated plant state according to switch position and the actual plant state.

In the distributed computer system the use of extensive software is permitted to achieve the most effective man–machine interface possible. This may reduce reliability, relative to the alarm facia system or discrepancy scheme, but the software involves extensive self-checking and other techniques. The intention is to ensure that the ultimate level of reliability is high relative to other systems of this type.

Ergonomics design principles

The main ergonomics design objective is to ensure that the alarm (and other interfaces) within each system provide only relevant, unambiguous and rapidly understood indications of the plant state. This is achieved via seven different principles:

- careful consideration of operational needs;
- careful classification and prioritization of alarms;
- clearly assigning and coding alarm 'ownership';
- ensuring that visible alarms are relevant to the task;
- using hardware and software logical alarm reduction;
- the application of clear audible and visual coding;
- ensuring compatibility of the man–machine interface between different alarm systems.

Ergonomics expertise has been applied throughout the design process to provide additional insights by means of systematic methods where required.

Practical implementation of ergonomics principles

The use of alarms to understand the state of a continuous process plant can be made difficult by the rapid onset of a large number of alarms; many of which are irrelevant to the situation in hand (Sheridan, 1981). At Sizewell 'B' the application of ergonomics principles makes it possible to reduce this number so that many of those alarms not relevant to main control room operations can be eliminated. How this is achieved is described below.

Operational needs

Within a pressurized water reactor the faults which are significant to safety are characterized by the onset of multi-system plant actuations invoked by automatic protection, a reactor trip and turbine trips, and attendant changes in the appearance of the interface.

The discrete and temporally displaced information provided by alarm thresholds is generally inappropriate for the purpose of scrutinizing changes in the continuous variables which characterize the changing state and performance of a pressurized water reactor. As described by Rasmussen (1974), alarm system information can only provide a coarse guide in identifying the nature of a fault. Alarms are, of course, well-suited to the initial indication of the presence of a fault. In a pressurized water reactor the distribution of alarms across several formats or subsystems is a characteristic of a safety-related fault, or other important change in the state of the plant. Much less reliance has, therefore, been placed upon the presence of alarms for procedurally driven diagnosis, and more upon the use of continuous variables, whose values differentiate faults.

Following the recognition of the presence of a major fault, characterized by a reactor trip demand or the actuation of safety protection systems, a single procedure will be entered for fault diagnosis.

In parallel with detailed fault diagnosis, critical safety function (CSF) monitoring will be undertaken. The supervisor will do this using bespoke formats on the distributed computer system, when available. Alternatively, the parametric high integrity VDU-based safety information display system and the alarm facia system can be used. This ensures that should the diagnosis or subsequent recovery prove inappropriate, the CSF monitoring will detect the consequent effects on nuclear safety and provide recommendations for recovery. The use of parametric data will diminish the importance of alarms for formal diagnostic purposes. Nevertheless, the visual pattern of alarms can create strong initial impressions on the nature of a major fault. It, therefore, remains important to provide a clear relevant and concise alarm system.

In the case of simple faults, a separate single procedure is provided to achieve diagnosis in response to alarms. Simple faults are those with no safety-related actuations and the onset of few alarms, such as those associated with single items of plant failing. In this case the alarms can be seen embedded within the system-based mimic formats so enabling a more effective diagnosis. Should sequence or onset time be an important issue, then conventional alarm list information can be consulted.

Alarm classification and prioritization

Alarms have been clearly defined from the start of the design process as 'those events on which the main control room crew must act'. This definition has been subdivided into 'priority alert' and 'alert'. All alarm facia system alarms are 'priority alert' whilst those on the distributed computer system, except alarm facia system replicas, are of 'alert' status.

Alarms not meeting those definitions, i.e. those not required by the main control room crew, but which are of relevance for maintenance, equipment condition monitoring or local operations, are consigned to other information systems for use elsewhere. This is now technically possible due to the ready commercial availability of effective network systems.

Alarm 'ownership'

All alarms annunciated within the main control room are either 'owned' by the operators or their supervisor. The plant is divided up into around 60 plant systems. Operators and supervisor have fixed allocations of system 'owner-ship' and operation within the overall plant. Within the distributed computer system all alarms can be viewed at any DCS workstation. Software flags ensure that alarms can only be accepted or reset at workstations used by the person 'owning' that alarm. This ensures that alarms are not removed by the action of others, but plant interactions between processes within the plant can still be readily understood by means of access to all distributed computer system formats.

The same allocation of 'ownership' by process exists within the discrepancy scheme system on the control consoles. However, the alarm facia system differs from the discrepancy scheme and the distributed computer system. Given the overwhelming importance to nuclear safety of the alarms within this system both the operators and the supervisor must examine the alarm facia system interface or its information replication within the distributed computer system.

Task relevance

Whilst prior classification ensures that only operationally relevant alarms appear on the main control room main–machine interface, relevance to the user at the time is entirely a function of the particular task in hand. Thus, the alarms relevant for notifying the onset of a fault are not necessarily those relevant to the tasks and subtasks used to achieve effective recovery or mit-igation of that fault.

The information system cannot by itself differentiate the user's stage of assimilation, recognition, diagnosis and recovery. Accommodating the differ-ences between these stages in the user's information needs has been achieved within the interface design in three main ways. Firstly, by the use of post-trip and fault-based formats for assimilation and recognition. Secondly, specific formats have been designed for use following major faults; these contain both parametric and alarm-based information. Thirdly, user involvement in the design has helped ensure that the embedding of alarms in system based mimics provides groupings relevant to task needs for major fault recovery.

It is anticipated that post-design, further reduction of alarms by reference to tasks may be possible via additional task analytical work on the Sizewell 'B' full-scope, high fidelity, simulator and by operational experience.

Logical reduction

This feature of the distributed computer system is being judiciously applied within the design process by reference to task needs. It is possible to apply hardwired, conditional and non-conditional logic at the plant, to remove the consequential alarms within a process following the loss of that process's prime function. Consequential alarms often report plant alignment status, not loss of system performance. Alignment information is more readily assimilable via the full use of the comprehensive set of distributed computer system mimic formats.

Alarm significance modification within the distributed computer system can remove alarms which are not relevant to the current plant operating mode. Plant operating modes are formally defined and universally recognized operating regimes for a pressurized water reactor. These modes are entered and left deliberately, and the operators are therefore well aware of their choice of current intended mode for the plant. The modes range from re-fuelling, through cold shutdown to hot, fully pressurized, at-power operation. In each of these modes different subsets of the full population of alarms signify abnormal conditions and are, therefore, unwanted and unexpected. The application of alarm significance logic can derate the alarm from 'alert' to information status only, when appropriate. (This ensures a fail-safe i.e. an alarm annunciates if alarm significance is not applied or fails). Information alarms are displayed on a separate list and in a subtle, less salient, way on the relevant mimic.

Transitional conditions exist between pressurized water reactor plant modes. For the operators, the point at which the transition from one mode to the next is made depends upon several criteria. These include the availability of safety equipment, the existence of sufficient margins to allow deviation on important parameters, and preparedness elsewhere for the changes of plant state that follow. When going from cold shutdown to power, reversals of mode progression may be necessary for reasons of safety; so it is for the user to determine when to change the alarm significance regime. The distributed computer system provides automatic detection and suggests the mode that may be chosen by the user. However, it is possible for the plant to 'run ahead' of the users' intentions. So it is appropriate for the operators to choose the moment for mode change in the alarm significance regime. This ensures that the operator decides what they expect and the system provides alarms accordingly. For example, an event which constitutes an alarm at-power may not be a matter for concern in any way at hot shutdown.

Invariant logic also allows the reduction of the alarm population by grouping several parameters within the distributed computer system. Two additional forms of alarm reduction exist on this system. These are the shelve and release functions. The shelve function allows the temporary removal of faulty alarms due to transmitter or other failures. The release function is used to remove alarms which are not of relevance in the plant condition but which cannot be reset due to the alarmed state persisting. This will only be used in circumstances where automatic alarm conditioning has not removed an alarm which is irrelevant to the current plant state or mode. As a disincentive to abuse, only one single alarm can be selected at a time from the interface and shelved or released. Further alarms requiring like treatment can then each be selected one by one and added to the shelved or released database. If the threshold which initiated the particular released alarm is recrossed as the parameter returns to the non-alarmed state, then the released alarm will be un-released. Thus, any further regression to the alarmable state is re-annunciated.

The simple software and the importance of alarm facia system alarms precludes the use of the shelve and release facilities.

Clear coding – audible alarm annunciation

Four audible alarms exist that are associated with the plant. On the distributed computer system two different audibles exist: one shared by the operators and one for the supervisor. Thus, task interruption is minimized for the non-recipient and the activity, which can be on a format other than that in view, will be sought by the relevant person 'owning' the audible heard. To assist in this, three overview screens in the main control room provide summary graphical alarms to indicate the format node(s).

In the case of the discrepancy scheme a single audible alarm is needed as light code activity within the switch can be seen clearly on the control panels from any point in the main control room. The light can be steady, or slow, or fast flashing according to the nature of the discrepancy being annunciated. The audible remains the same but is sufficient to draw attention to the interface. Acceptance is signified by the manual changing of switch position. This will cancel an audible not already cancelled (provided other discrepancies do not exist). Because 'ownership' of the alarm facia system is shared by all and requires universal attention only one audible exists for it.

The alarm facia system, discrepancy scheme and distributed computer system share a single function to cancel an audible alarm from any one or more of them. The annunciation of audible alarms are time-sequenced such that a distributed computer system operators' alarm precedes a DCS supervisor's alarm, which precedes a discrepancy scheme alarm. Only the alarm facia system audible alarm annunciates in parallel with the other three. Their relative perceived audible importance/urgency has been designed using rhythmical and chordal structures, and were determined to be acceptable and correctly perceived, by experiment.

Clear coding – visual annunciation

Alarms are differentiated by spatial location and the appearance of the interface within the control room except within the distributed computer system. Here clear visual codes have been used to differentiate 'priority alert' from 'alert' alarms. The alarms have different marker shapes with redundant colour codes; a red triangle for 'priority alert' and an orange square for 'alert'. The hierarchical location within the formats shows 'ownership'.

Within the distributed computer system the area of plant in which alarm activity exists is also clearly summarized by means of the permanently displayed plant overview format, which provides an alarm cue for each plant system with one or more alarms. This cue also indicates whether the alarm has priority or alert status.

Inter-system interface compatibility

The distributed computer system and alarm facia system have the same sequential stages of alarm handling. These are:

- audible cancel–signifying awareness of the existence of an alarm;
- accept–signifying the alarm has been read and understood; and
- reset – which allows all trace of the alarm to be removed once the plant condition has reverted to normal.

On the discrepancy scheme alarms are both accepted and reset by switch action unless the item of plant has a persistent fault.

Use of the alarm systems in practice

When the alarm systems are all working the handling and use of alarms will be done using the distributed computer system. The audible alarm can be separately cancelled and alarm facia system and discrepancy scheme visual cues may be left displayed until time allows for their handling.

Within the distributed computer system many faults in the transmission of individual parameters are automatically trapped and displayed as invalid by software. When they are not, they may be detected by the user noting an incongruous depiction of the state of the VDU-depicted system. This incongruity can explain the existence of a single alarm associated with a continuously variable parameter. Normally the existence of one, out–of–normal, variable will be accompanied by the deviation of other coupled variables.

Additionally, routine checks will be undertaken between the distributed computer system and safety information display system to ensure agreement of parameters between the two. The time for which disagreement can be tolerated between the two systems without further action will be dictated by the station's technical specifications. This time will vary according to whether or not the plant is in a steady state at the time. (It should not be forgotten that unwished-for conditions are more likely to arise when the reactor is tripped; so introducing a transient into an otherwise steady process.)

Should the distributed computer system suffer a systemic failure this will be detected by self-checking software or by the user noting the freezing of the dynamic screen cue. Thus the integrity of the alarms can be maintained by automatic and user checking, and by the judicious and controlled use of the shelve and release facilities on the distributed computer system.

Future potential improvements in alarm systems

Outlined below are some possible means to reduce the cognitive burden of dealing with alarms. In essence the handling of alarms can potentially become an operational end in its own right; it is important that attention is not held at the alarm interface when other diagnostic or control tasks need to be performed.

Not all the enhancements suggested require more sophisticated technology, but do acknowledge the need to make alarm annunciation and handling compatible with the user's needs. Despite the size and power of modern distributed computer systems it is important to note that the computational operations associated with alarm management and display still form a substantial portion of its workload. It, therefore, remains important to resist the excessive use of this still limited capacity.

Artificial intelligence (AI)

It is generally expected that the use of artificial intelligence (AI) in the form of an expert system will offer benefits to the user in providing automated recognition of faults within a process plant. To date the major application of AI in the nuclear industries has been for planning in off-line systems (Tate, 1986). In practice, the application of such solutions requires extensive design lead-time and analytical effort to achieve a mathematically accurate plant model which is never misleading to the operator, and also to ensure that the software is itself reliable. Software qualification would also be even more challenging than for current systems. The vulnerability of such systems to parameter measurement failure and consequent failure in diagnosis can be significant and should not be overlooked.

A paradox in the use of such technology is apparent in the domain of reliability. That is, how to trade-off the potential improvements in human diagnostic reliability against the decreased reliability (or at least, the increased difficulty in qualifying the software) of the information system. In addition any limitations in the scope of such automated diagnosis could lead to a reduction in the user's diagnostic skills. This could be either because they are less frequently used or because of an inappropriate allocation of diagnostic tasks between user and technology. In the intermediate term improvements for high integrity systems may lie in the enhancement of existing alarm systems interfaces.

Alarm relationships with tasks

As stated above, the set of alarms required is a function of the plant operating mode, the user's task in hand and their expectations. The Sizewell 'B' system interfaces are designed with the explicit recognition of task needs.

However, the extent to which these interfaces are able to match the task needs is currently limited by technical constraints common to all process monitoring systems.

For example, a given physical parameter on the plant may require to be alarmed at a number of thresholds relevant to different tasks. Within a given direction of deviation from the norm, six thresholds may be needed to indicate action as follows:

- the need to make fine adjustment;
- avoidance/preparation for an automated trip;
- reinforcement/confirmation of an automated trip;
- the avoidance of breach of an acceptable safety limit;
- recovery from transgression of that safety limit.

This number of thresholds could well be doubled in the most extreme cases for deviations in the opposite direction, to give a total of 12 alarm thresholds. The majority of current alarm systems can only support 4–6 thresholds per parameter. (This issue is currently addressed by the use of multiple transmitters for different range bands, which limits the extent to which less extreme and irrelevant thresholds can be suppressed). Further modifications to these thresholds may be needed to ensure that the alarm annunciates at a parametric level relevant to the operational mode of the plant.

Such extended alarm information could, of course, only be fully used within relevant task-based mimic formats. Although user demand would also lead to the need for compact coding of all thresholds within system-based formats. Clearly the number of thresholds passed challenges the potential usefulness of single alarm lists and suggests the need for different types of alarm lists sorted on differing criteria to limit the number of alarms per list.

Additionally, a single threshold may be relevant to more than one user or format and handling may be improved by having separate handling for each format when needed. Current systems often do not meet this need.

Literal and derived sensing

Another engineering issue centres around the forms of sensor/transmitter used to convey alarmable information into the interface serving system. For example, it is conventional to detect the state of an electrically-driven pump by the use of circuit breaker position detection. However, this form of detection cannot confirm whether the impeller is rotating or not during mechanical failures within the pump. The situation could be avoided by the development of reliable transmitters which directly sense the state of concern – an important factor for both alarms and plant monitoring purposes.

Diagnostic templates

The preceding interface enhancements will largely stem from the technical ability of information systems to reliably sustain a larger amount of data and calculations than hitherto.

Within the domain of ergonomics the improvements, which can be achieved concomitant with the engineering improvements, are likely to stem from enhancement of the user's ability to specify the nature of the task in hand and hence select more forms of display. Shallow-wide, system-based hierarchies will probably still form the backbone of information systems, largely because of user familiarity and ease of navigation. Within formats, the visual existence of more alarm thresholds could however be controlled by the use of information masking. This would require the information user to inform the system which particular task was under way and accordingly, alarms would be selectively revealed.

The notion underlying this simple enhancement stems from the fact that alarms are there to announce the unexpected but important changes of plant state. The user only wishes to be informed of the unexpected. Unfortunately the system cannot 'know' what the user expects at any given time and must therefore, perforce, display all information which *might* be unexpected. Perhaps the user can use templates or otherwise describe to the system what is and is not expected (see Usher, Chapter 8).

It may be possible to exploit human pattern matching capabilities further by the use of tabular templates. These could each display a pattern for a set of parameters which conform to a given significant fault and which also shows the current values. This would further extend the principle of parametric diagnosis which is to be used in the Sizewell 'B' main control room and reduce the onus placed upon alarm systems for the diagnosis of multi-system faults.

Multi-dimensional displays

Another promising variation in the display of information is illustrated in the approach put forward by Beltracchi (1987). Here, the emphasis moves from a focus upon the condition of the plant in terms of system alignments and parametric measures of state, to one in which the thermodynamic conditions are displayed. Specifically, a format which depicts a mimic of the secondary side of the plant (steam generators to turbines and back again) displays the thermodynamic temperature gradients and the parameters which relate to the Rankine cycle. (The Rankine cycle provides the thermodynamic model of heat transfer from reactor to turbine power). Any faults in the heat sources (reactor and reheaters and feedheaters) or the heat sinks (turbine or condensers) leads to a resultant distortion in the mimic diagram. This multi-dimensional display clearly reflects many of the faults of interest to both safety and thermal efficiency. It appears to be well accepted by users. Otherwise, to date, multi-dimensional displays have not been well received by designers and operators alike. Beltracchi's success probably stems from the face validity of the display which shows a recognizable representation of the plant rather than some abstract multi-dimensional depiction such as a smiling/frowning face.

Automatic control and protection

Design and ergonomics creativity remain challenged by the user's informational needs for the detection of abnormal states in automatic control and protection systems. The need for manual reinforcement of a protection system failure is beyond the design basis for Sizewell 'B'. This is due to design diversity and redundancy of these systems. However, the user needs information on the state of the protection systems in order to check their operation. This can be done during testing and prior to shutting down superfluous safety equipment following a fault. The need for their personal reassurance of correct operation is also important, irrespective of the design basis. Several approaches to the display of such systems have been attempted in the past. In the view of the author none of them has received high acceptance from users. The depiction difficulties stem from both the logical complexity, the breadth of such systems and the fact their operation is an essentially abstract phenomenon which appears to be held by the user as mental models that are a combination of cause–effect, temporal sequence and are heuristic in their form.

Reinartz (1993) has found that on German pressurized water reactors the display of *which* actuating action or interlock, *what* event or state has triggered the automatic system and *why* the actions have been activated, are powerful aids to understanding the actuation of automatic protection systems. It is an interesting fact that 'reasons why' are rarely displayed explicitly within the information which is the interface, but are left to training.

Alarm prioritization

Prioritization of alarms remains an issue where future improvements in design solutions are conceivable. There are four factors which affect the significance of an alarm to the user:

1. the time scale on which the alarm must be read and acted upon;
2. the importance of the safety consequences of taking no action;
3. the importance of the productivity consequences of taking no action; and
4. the relevance of the alarm to the task in hand.

It is inconceivable that a single dimension of alarm importance may be satisfactorily derived which reliably takes account of all four factors. It is, however, possible that the concept of CSFs, which has clearly identified the most important plant parameters and their attendant alarms, could be extended to the identification of critical availability parameters. Similarly Beltracchi's concept may be extended to depict other relevant fundamental processes which can be expressed by physical models and displayed multi-dimensionally. Thus ensuring that the most threatening alarms are quickly assimilated.

Instant and paged access

The use of a separate interface such as the alarm facia system can make rapid assimilation and recognition more certain as they do not require inter-format navigation and paged-access. Also, with careful design and a limited family of alarms, clear patterns differentiating major faults may be possible if coupled with limited computational power. This would be used for the calculation of derived variables not obtainable directly and which differentiate faults. Beyond that, hidden software flags and visible multi-dimensional tagging codes within VDU displays may make the sorting of alarms by the four different factors possible. Thereby enabling the user greater choice in the criteria applied to the display of current alarms.

Alarm cues and action cues

Audible alarms have valid uses which are not a matter of alarm. These include the indication that an anticipated action is now possible. For example, an action that can only start when another has been completed. This could be a unique audible, assigned by the user with a value and parameter of their choosing. Further clarification may be attained by the differentiation of cues used to confirm the successful completion of an ongoing automatic action or slow process evolution as opposed to the existence of the unexpected fault. Within Sizewell 'B' such cues do not currently have an audible or visible annunciation as alarms but in systems elsewhere may be treated as alarms.

Extended processing power and careful use of audible displays developed in the ways described by Edworthy (chapter 2), does make such an approach conceivable in the future.

Information system capacity

Each of these ways to expand the forms of user interface will demand greater processing and data capacity than is currently available if data update times are to remain acceptable. Nevertheless the utility of such forms of interface should be explored.

Conclusions

The early recognition within the Sizewell 'B' design process of the important engineering and ergonomics principles, together with comprehensive ergonomics user and designer attention to detail, should achieve an effective, compatible and integrated man–machine interface. Many operational problems surrounding the use of alarms will have been much reduced, relative to many other plants.

In future, the use of AI-based systems to provide alarms to assist in diagnosis raises the prospect of diagnostic de-skilling and poor system reliability, which could lead to reduced man–machine system reliability. Further improvements in instrumentation and alarm threshold software may lead to clearer alarms relying less on user inference. The creative use of radical display techniques may also enhance the user's abilities by providing more choices in the manner of alarm depiction for particular tasks.

Acknowledgements

Thanks are given to the Nuclear Electric PWR (pressurized water reactor) Project Group for permission to publish this paper. The views expressed in this chapter are those of the author alone.

References

Beltracchi, L., 1987, A direct manipulation interface for heat engines based on the Rankine cycle, *IEEE Transactions on Systems, Man and Cybernetics*, **17** (3), pp. 478–87.

Corcoran, W.R. *et al.*, 1981, The critical safety functions and plant operation, *Nuclear Technology*, **55**, 689–717.

Edworthy, J., Urgency mapping in auditory warning signals. In this book.

Rasmussen, J., 1974, Communication between operators and instrumentation, in Edwards, E. and Lees, F.B. (Eds) *The Human Operator in Process Control*, London: Taylor & Francis.

Reinartz, S.J., 1993, Information requirements to support operator-automatic co-operation, in the *Proceedings of Human Factors in Nuclear Safety*, Le Meridien Hotel, London.

Sheridan, T.B., 1981, Understanding human error and aiding human diagnostic behaviour in NPPs, in Rasmussen, J. and Rouse, W.R., *Human Detection and Diagnosis of System Failures*, pp. 19–37, London: Plenum Press.

Tate, A., 1986, *Knowledge-based Planning Systems*, in Mamdani, A. and Efstathiou, J. (Eds) *Expert Systems and Optimisation in Process Control*, Technical Press.

Usher, D., The Alarm Matrix, in this book.

US Nuclear Regulatory Commission, 1981, Functional criteria for emergency response facilities, Palo Alto, NRC NUREG 0696, p. 37.

Part 4
Applications of alarm systems

Applications of alarm systems

Neville Stanton

This section presents three chapters based on the different applications of alarm systems. Chapter 11 (by Ed Marshall and Sue Baker) discusses alarm systems in nuclear power plant control rooms and (briefly) air traffic control. Ed and Sue suggest that despite the wide range of applications of alarm displays in process control (e.g. nuclear power, conventional power, chemical plants, oil and gas production) the problems faced by operators are generally very similar. Traditionally alarms were presented on illuminated tiles (annunciators) but the advent of information technology has enabled the alarm information to be presented on VDUs. Ed and Sue question the rationale of this move, to suggest that alarm lists may not be the most effective manner to support the operator in managing the process. They propose that there are a number of strategies that could be employed to improve alarm presentation, such as:

- a return to annunciators;
- alarm filtering;
- alarm logic systems;
- function-based warning systems;
- diagnostic expert systems.

Finally, Ed and Sue consider how operators use alarm information in their daily activities. This provides a qualitative insight into the demands the alarm system makes upon operators.

Chapter 12 (by Chris Baber) presents in-car warning devices. Chris focuses on the psychological issues involved in designing and implementing alarms in cars. He starts by considering the basic issues involved with the presentation of information to drivers, such as with non-alarm information, e.g. speedometers. Chris then provides a classification of displayed information, namely:

- tell-tales (devices that remind drivers of status, e.g. indicators on);
- advisories (devices that inform drivers of potential problems, e.g. fuel low);

- warnings (devices that alert drivers to immediate problems, e.g. temperature high).

Finally, Chris considers novel formats for in-car alarm displays, such as reconfigurable, auditory and speech displays.

Chapter 13 (by Tina Meredith and Judy Edworthy) discusses alarms in intensive therapy units. They note that many of the problems associated with auditory alarms in hospital environments are psychological in nature, i.e.:

- it is difficult to learn more than six or seven;
- they are often confusing;
- they are often inappropriate in terms of urgency mapping.

In order to elicit a clearer picture of auditory confusion in intensive therapy units Tina and Judy conducted a series of experiments where they presented subjects with natural alarm sounds, controlled alarm sounds and controlled, neutral names, alarm sounds. Their results showed that natural alarm sounds are more often identified correctly than the controlled alarm sounds. Tina and Judy suggest that this has a clear practical implication for working practice: a relatively small amount of time spent training the nursing staff in the different auditory alarms would pay dividends in terms of alarm recognition.

11

Alarms in nuclear power plant control rooms: current approaches and future design

Edward Marshall and Dr Sue Baker

Introduction

For the control room operator in a power plant, it is generally assumed that the first indication of a process disturbance will be the alarm system. The aim of this paper is to illustrate the alarm types currently in operation in power plants and to chart the research and development of new alarm systems now proposed for power plant control rooms. The first section of this paper concentrates on both alarm systems in nuclear plants and developments taking place in the nuclear industry, this is because most research has focused in this area. However, it should be remembered that alarm displays, and the problems faced by the operator using them, are very much the same in fossil-fired power stations, chemical factories, and oil or gas production plants. By drawing on examples of ways that operators use alarm systems during the day-to-day operation of the plant, the second part of the paper suggests that screen-based alarm systems may not satisfactorily support routine tasks involving the alarm system because of the design emphasis on dealing with serious plant disturbances.

The intent of the paper is not to discuss psychological models in any detail but to address practical issues observed by the authors, principally in the power industry but also in general process control and aviation, where safety can depend on the proper use of alarm systems.

The role of the operator

Skill in process operation has developed in parallel with technological developments which have typically involved a steadily increasing degree of automation. Modern, large scale process plants, nuclear or conventional, whether for power or chemical production, could not be operated without automation and computers have been exploited in plant control since the early sixties. However, even with this high level of automation, the operator is still very much involved in the day-to-day running of the plant. Often automatic systems will not take over control until steady running conditions have been established. Thus during startup and shutdown there is still heavy reliance on manual operation. Even at steady, full power operation, the operator will continually monitor and trim the plant to maintain efficient running conditions. Most importantly, the operator is responsible for the early recognition and diagnosis of fault conditions. It is the alarm system which principally supports this crucial element of the operator's role.

Alarm presentation

In a nuclear plant control room alarms are important and thus, distinct visual signals are displayed to attract the operator's attention to abnormal plant conditions. It should be noted that onset of an alarm is usually coupled with an auditory signal (typically, a chime or tone) and the first action of the operator will be to silence it before attending to the content of the message. Two modes of alarm presentation are to be found in today's power plant control rooms.

1. Traditionally, alarms have been presented as messages on illuminated tiles known as annunciators which light up to show an inscribed message, e.g. 'high level in pressurizer' or 'low temperature in superheater'. The operator is able to use pattern recognition and geographical cues as an aid to diagnosis. However, in a rush of alarms, the operator has little chance to follow the development of the disturbance over time.
2. The advent of computer technology has enabled the presentation of alarm information on VDU screens. Such displays currently tend to be in the form of tabular lists of single line text messages presented in time order. A new alarm will appear at the bottom of the current list, and as the screen fills after 20 or so messages, the earlier messages scroll off the top of the screen. Facilities may be provided for paging back and forth through the list and various coding techniques (colour and symbols) may be provided to distinguish message priority. The operator thus has powerful cues as to time sequence but little in the way of patterns or location to assist in diagnosis. In modern UK nuclear power station control rooms the majority of alarms are now presented on VDU screens. For example, at Heysham 2, a recently commissioned UK advanced gas-cooled reactor (AGR), there are about 18 000 alarms per reactor (Jackson, 1988).

Alarms for fault identification

The current development of alarm systems can be considered within the framework of an idealized and simplistic three-stage decision model which describes how operators cope with process faults:

1. Detection – the operator needs to be able to detect the onset of a plant disturbance.
2. Diagnosis – the operator should diagnose the particular disturbance from the presented symptoms.
3. Remedial actions – the operator should select and carry out the appropriate remedial action to mitigate the disturbance.

Clearly the conventional alarm panel can effectively alert an operator to the occurrence of a plant fault, but the sheer number of alarms activated, and the fact that different faults can produce similar alarm signatures, render correct diagnosis difficult. On his side, in this otherwise potentially confusing array of information, is a slow acting dynamic process where automatic safety systems have been installed to bring about a safe and controlled shutdown. The rapid onset of large numbers of alarms during a plant disturbance can be a serious problem for the operator trying to assimilate all the information for performing a rapid and accurate diagnosis of process state. Increased concern over the design and operation of computer-based alarm systems was raised by the Three Mile Island (TMI) incident in the USA in 1979. Then nuclear power station control room operators were overwhelmed by the sheer number of alarms and, in fact, this was cited as the main reason why operators overlooked crucial alarms.

Improvements in alarm presentation

A number of alarm handling features are already implemented in control room information systems to assist the operator. Examples are prioritization of alarm messages, indication of *first up* and *prime cause* alarms and the facility to shelve *nuisance* alarms. Such features can assist in diagnosis by directing the operator towards the most relevant alarm messages.

Nevertheless, this flood of information can be unintelligible and confusing to the operator, particularly if he/she must interpret it and take action within a short time. Two broad techniques have been suggested to improve the diagnostic potential of alarm systems.

1. Application of logical filtering to remove all but essential messages (e.g. present only alarms which require action by the operator); and
2. improvement in the presentation of alarm information.

The principal concerns relate to the more general aspects of alarm displays – whether annunciator tiles are more effective than VDU text lists, and how computerized information can be enhanced by means of display hierarchies, plant overviews and graphic mimic diagrams.

Research examples

A number of experiments have been carried out in recent years, involving various degrees of simulation and different subject populations, with the object of exploring performance aspects of alarm presentation.

The US Electrical Power Research Institute (EPRI, 1988) has reported a study in which a small group of TMI operators were observed using five different alarm systems. The alarm systems were simulated within the context of a mockup of the control room. The five different systems comprised three different arrangements of conventional tiles and two different text-based VDU alarm lists. Performance was generally better with the annunciators, and operators expressed a preference for this system. The finding is, however, probably not surprising given that the subjects were familiar with annunciators and that the VDU displays were limited in the way information was presented, i.e. limited use of colour and no use of graphics.

A dynamic alarm filtering system developed for Japanese pressurized water reactors (Fujita, 1989) has recently been evaluated in an experiment in which nine reactor crews demonstrated 40 per cent faster fault detection times than when using a conventional system.

The OECD Halden Project in Norway has carried out a series of experiments on alarm handling in nuclear plants. These were aimed at evaluating their HALO (Handling of Alarms using LOgic) system which combines dynamic on-line alarm filtering in terms of plant status together with advanced graphic techniques for the presentation of alarm information. In an experimental evaluation, 10 operators used either HALO or an array of conventional alarm tiles to handle a complex transient scenario. The results showed clear performance advantages, in terms of diagnostic accuracy and response time, when operators used the advanced graphic based alarm presentation system (Reiersen, Baker *et al.*, 1988).

Function-based warning systems

An alternative strategy to presenting the operator with specific diagnostic advice is to provide information regarding the status of higher level plant functions.

The safety parameter display system (SPDS) was a post TMI enhancement specified by the US Nuclear Regulatory Commission for backfitting to existing control rooms. SPDS is intended to extract and present a set of key plant parameters on a dedicated panel or VDU in order to enable the operator to carry out the actions required to maintain the plant in a safe condition.

Combustion Engineering Inc., for example, have developed a sophisticated SPDS – the critical function monitoring system (CFMS) – which is based on the assumption that the safety state of a process can be expressed in terms of a small number of critical safety functions. As applied to a nuclear power plant, a critical safety function is defined as:

> a function that must be accomplished by means of a group of actions that prevent core melt or minimize radiation releases to the general public.
>
> Corcoran, Finnicum *et al.* (1980)

Thus, during a complex disturbance, while it may be difficult for the operator to locate and diagnose the precise cause, by maintaining critical functions the plant can be kept in a safe condition. In the CFMS, computer algorithms use several hundred plant signals to express the status of each critical function.

If a critical function is threatened the operator carries out remedial actions to relieve the threat to the critical function without necessarily making a precise diagnosis of the disturbance. The remedial actions are presented as success paths which are defined as the actions associated with the automatic or manual deployment of plant components and/or systems which can be used in restoring or controlling a critical safety function.

A number of experimental studies were carried out during the 1980s to evaluate prototype SPDSs (Woods, Wise *et al.*, 1982; Hollnagel, Hunt *et al.*, 1984). These involved the observation of operators using SPDSs to cope with transients implemented on training simulators. Although clear statistical evidence in favour of the systems was not obtained, operators expressed strongly positive opinions for their inclusion in control room enhancement plans.

As a result of these experiments, the CFMS has been enhanced to provide additional information to the operator on the availability and suitability of various remedial operations which could recover the plant situation (Marshall, Baker *et al.*, 1988).

Such functional based alarm systems are now installed in modern nuclear power plants. However, they have yet to find application in fossil-fired power stations or petrochemical plants.

Diagnostic expert systems

Expert system designers are attempting to encapsulate the skilled operator's diagnostic skill within a computer programme. Early attempts tended to be slow and had difficulty in dealing with faults not previously programmed. More recently the advent of artificial intelligence programming techniques has provided a promising vehicle for the development of systems capable of fast versatile diagnosis, though these have yet to be tested in a real plant environment. An example of a system which uses patterns of alarms and other salient plant parameters to propose possible diagnoses is described by Berg, Grini *et al.* (1987).

The day-to-day use of alarm systems in plant control

The sudden onset of a serious disturbance at some time during a period of steady running with otherwise stable plant conditions, has provided the basis for the design of the alarm handling systems described above. As already mentioned, the Three Mile Island incident is the most often cited seminal example (Kemeny, 1979). The basic issue is that although the sequence of alarm signals may be completely novel to the operator, he must, nevertheless, interpret them so that corrective action can be put in hand. Although efficient tools for responding to the unlikely, and thus fortunately rare, process faults are necessary, designers should not overlook the fact that the alarm system is continually in use during day-to-day plant operation. Anyone who has spent time in the control room of a power station or chemical plant will have noticed that alarms are a constant background to the business of operation. It could be argued that the stereotype of the systematic use of alarms for process fault identification, as proposed by cognitive task analysts, where alarm onset triggers a sequence of increasingly complex mental activities, (Rasmussen, 1986) does not at all reflect the typical use of control room alarms.

Active monitoring

The alarm system is actively monitored during normal running and patterns of lamps on the alarm panel are exploited continually to confirm plant component status. The traditional alarm fascia thus provides the operator with a richly detailed information source for ascertaining plant status at any time during a complex process manœuvre. Take, for example, the task of running up a coal-fired power plant. The procedure entails sequential starting of six to eight individual coal mills. In the traditional control room each mill has its own alarm fascia. Since the operator is familiar from long experience with the pattern of lamps corresponding to shutdown, or normal running, a single glance at the fascia allows him to check on the condition of each mill. Any discrepant pattern immediately alerts him to possible problems and the location of the signal may immediately direct him to suitable remedial action. Screen-based alarms may well hold the same information but, by requiring active searching on the part of the operator, they do not yet provide the same degree of support for the task.

Alarms as a source of reassurance

Operators expect the occurrence of alarms as they manœuvre the plant and this expectation is matched against the activation of alarm signals. Consider again a coal-fired power station but, in this case, plant shutdown. As the plant· cools, alarms are periodically triggered as temperatures and pressures fall below their normal running values. The designer of computerized alarm systems would seek to eliminate these alarms, considering them as a nuisance. However, operators use them to check that the plant is behaving normally and as reassurance that the shutdown is progressing at the right pace. When the alarm audible signal occurs, the operator glances at the expected fascia, sees that the alarm corresponds with his expectation of plant status and simply acknowledges it. The alarm may be anticipated but it is not a nuisance.

Fault identification

Obviously alarms alert the operator to faults, but the recognition of an aberrant alarm is embedded in the operator's activities in relation to the desired plant condition. The operator's awareness of plant status can render a single glance at the alarm fascia sufficient for the identification of an out-of-context alarm, the diagnosis of its cause and the fluent implementation of corrective action. An alarm system which, after the initial alert, demands a number of sequential search actions through several screens of information is a potential source of disruption to this skilled performance, not only is this irritating but it may well lead to inefficiency in coping with the more routine plant incidents.

The role of alarms in air traffic control

In a discussion of the functions of the alarm system, it is perhaps relevant to consider a task outside process control. The air traffic controller task is likened to the plant operator principally because in both situations small groups of people exercise centralized control over remote processes (Reason, 1992; Sanderson, Haskell *et al.*, 1992). Though these similarities are undeniable, there are also critical differences in these tasks (Baker, 1992). These differences mean that design philosophies for key components of the interface, like the alarm system, could be inappropriate if simply transferred from one task domain to the other.

The air traffic controller is unlikely to face precisely the kind of disturbance which has been described in relation to plant operations. In air traffic control the problem leading to an alarm may well have developed as a function of the way the controller has managed the situation. The occurrence of a conflict alert alarm is inherently diagnostic, i.e. it alerts the controller to a loss of separation between aircraft. Furthermore, remedial action will be obvious and well drilled. Clearly, an effective aircraft conflict alert warning will consist of an alerting signal coupled with a cue embedded in the working screen. A text message presented on a separate screen announcing a detected potential conflict, or a hierarchical system requiring a chain of searching, would be totally inappropriate given the spatial nature of the task and the need for a fast accurate response. For the designer of plant alarm systems, the air traffic control task serves to underline the value of spatial information in fault management.

Conflict alert is thus a short term and predictive alarm presentation system where the message is inherently diagnostic. The process plant alarm has a longer term response. It is generally warning of a past event and usually requires interpretation with subsequent diagnosis. Nevertheless they share a common problem in the matter of nuisance alarms, i.e. the occurrence of alarm signals when the operator is well aware of the disturbed plant state or of the conflicting track of two aircraft. As discussed above, nuisance alarms can play a role in the management of plant state change in keeping the operator in touch with the effects of process manipulation during a complex situation. They can still, however, provide annoying and potentially confusing disruption during the operator's task.

Conclusions

Alarm systems clearly give operators adequate indication of the onset of a plant disturbance. However, in their current form it is still up to the operator to diagnose the root cause of any alarms.

Developments in on-line logical filtering of alarms, function based alarm systems and the use of computerized graphics for alarm presentation are now well researched but they are not yet common in control rooms, although systems are specified for plants now under construction. Such enhancements have been shown by experiment to improve diagnostic accuracy and have been well received by operators. For the future, expert systems have been devised to propose diagnoses on the basis of analysis of plant alarms and to suggest appropriate remedial strategies. In this way alarm systems could soon provide support for the operator in all stages of dealing with serious and unusual process faults.

Thus, computerization is having a profound effect on the way in which process alarms are presented. These changes are directed towards assisting the operator diagnose serious and unexpected plant faults. There does seem to be a possibility, however, that designers are neglecting the way in which alarm systems are currently used by operators in the normal day-to-day running of the plant. These more mundane applications should be considered, in particular the way in which routine tasks are supported by the spatially arranged information on the traditional fascia panel. The problem of separating nuisance from useful alarms still has not been addressed satisfactorily; simply suppressing alarms on the basis of plant status does not reflect the way in which operators use them. Nevertheless, a degree of filtering is essential in reducing the sheer quantity of alarm information that may occur during a severe transient. The designer must seek to support the diagnosis of plant disturbances while preserving the role of the alarm system in maintaining the operator's overall view of plant status.

References

Baker, S.M., 1992, 'The role of human factors in the investigation of ATC incidents,' presentation at the Spring Seminar of the Safety and Reliability Society entitled Human Factors in Safety & Reliability – Contrasts within Industries, Altrincham, Cheshire, May.

Berg, Ø., Grini, R.E. and Yokobayashi, M., 1987, Early fault detection and diagnosis for nuclear power plants, *Atomkernenergie-Kerntechnik*, **50**.

Corcoran, W.R., Finnicum, D.J., Hubbard, F.R., Musick, C.R. and Walzer, P.F., 1980,

The operator's role and safety functions, AIF Workshop on Licensing and Technical Issues – Post TMI, Washington DC, March *C–E Publication TIS–6555A*.

EPRI, 1988, An evaluation of alternative power plant alarm presentations, *EPRI NP–5693Ps*, Vols. 1 and 2, Palo Alto, USA.

Fujita, Y., 1989, Improved annunciator system for Japanese pressurized water reactors, *Nuclear Safety*, **30** (2), 209–21.

Hollnagel, E., Hunt, G. and Marshall, E.C., 1984, The experimental validation of the critical function monitoring system: executive summary, OECD Halden Reactor Project, *HPR–312*, Halden, Norway.

Jackson, A.R.G., 1988, 'The use of operator surveys by the CEGB to evaluate nuclear control room design and initiatives in the design of alarm systems and control room procedures,' presentation at IEEE Fourth Conference on Human Factors and Power Plants, Monterey, California, June 5–9.

Kemeny, J.G., 1979, Report of the President's commission on the accident at Three Mile Island, Pergamon Press.

Marshall, E.C., Baker, S.M., Reiersen, C.S., Øwre, F. and Gaudio, P.J., 1988, The experimental evaluation of the success path monitoring system, *IEEE Fourth Conference on Human Factors and Power Plants*, Monterey, California, June 5–9.

Rasmussen, J., 1986, *Information Processing and Human–Machine Interaction: An Approach to Cognitive Engineering*, New York: North Holland.

Reason, J., 1992, 'The identification of latent organizational failures in complex systems,' presentation at the NATO Advanced Study Institute: Verification and Validation of Complex and Integrated Human–Machine Systems, Vimiero, Portugal, July.

Reiersen, C.S., Baker, S.M. and Marshall, E.C., 1988, An experimental evaluation of an advanced alarm system for a nuclear power plant – a comparative study, in Partrick, J. and Duncan, K. (Eds), *Training, Human Decision Making and Control*, North Holland: Elsevier.

Sanderson, P.M., Haskell, I. and Flach, J.M., 1992, The complex role of perceptual organization in visual display design theory, *Ergonomics*, **35** (10).

Woods, D.D., Wise, J.A. and Hanes, L.F., 1982, Evaluation of safety parameter display concept, Vols. 1 and 2. *EPRI NP–2239*, February.

12

Psychological aspects of conventional in-car warning devices

C. Baber

Introduction

This chapter considers the relatively mundane world of information display in cars in order to raise and discuss some points associated with the psychological aspects of 'alarm' use. The term 'alarm' is given a very broad meaning here in order to introduce some of the principles of psychology which can be assumed to be at play in understanding and interpreting alarm information. The domain of information displays in cars has been chosen because, probably, it will be more familiar to many readers than the process control or medical fields discussed in other chapters. Drivers may feel that car dashboards are relatively easy to understand and that the information presented does not provide too many problems. However, there has been growing concern regarding the possible safety issues surrounding evermore complicated instrumentation in cars; will there come a time when the complexity of car dashboards will lead to drivers spending more time looking at the dashboard rather than at the road? By considering the psychological processes involved in using in-car displays, predictions can be made about some of the possible problems associated with their use, and by extrapolation, potential problems with 'alarm systems' in other domains can be considered.

Instrumentation in cars

There has been a rash of research into 'high tech' applications of computer technology in cars. Air-conditioning used to be a simple matter of raising or lowering temperature; now it is possible to define a total thermal environment for a car. Trip computers and complex in-car entertainment systems have become commonplace, especially in luxury cars. Navigation aids, although still at the 'concept stage' for many manufacturers, are soon to be a reality.

It can be assumed, from looking at early models of road vehicles, that the basic idea behind in-car warning technology was to provide the driver with information which was not directly perceptible via his/her senses. Thus, there are indicators relating to the operational state of the vehicle, e.g. fuel level, oil pressure, water temperature, and to the performance of the vehicle, e.g. speed. This information was typically displayed either in the form of dials or in the form of coloured lights. Over the past 70 years we have seen advances in automobile capability and a corresponding increase in available information. Where a production car of the 1930s may have informed the driver of the state of half-a-dozen variables at most, a modern car can easily present in excess of 20 and future displays could easily present '. . . dozens of different engine, drivetrain and aerodynamic measurements' (Train, 1987). This is often simply due to an increase in sensing capabilities; it is now possible to provide accurate sensing devices on a wide range of automobile components. This raises the question of how far is this 'information explosion' warranted?

Knoll, Heintz *et al*. (1987) proposed a checklist of ergonomic factors which need to be addressed in the design of in-car information systems:

- minimum distraction of the driver;
- readily accessible operating elements;
- easily readable displays and inscriptions;
- speedy familiarization;
- minimal preknowledge;
- space saving dimensions;
- attainability with justifiable outlay using available technologies.

Although the Industrial Ergonomics Group at the University of Birmingham would not argue with these aims, we would note that it is also desirable to ensure a good match between drivers' requirements and what the technology provides. Figure 12.1 presents a simple diagram of information display requirements in cars (adapted from Simmonds, 1983). While one can see that 'warnings' form a subset of information types, Figure 12.1 shows the difficulty of considering 'warnings' in isolation from other information types, i.e. there is a limited number of information presentation formats which can be used in the car, and these formats are often used to present different types of information to drivers. The final column introduces some of the psychological processes which we can assume the driver engages in.

In the following section, the design and use of speedometers will be discussed. While this may represent a departure from the notion of in-car warnings to some readers, it is important to consider *how* information displays are used by drivers. The speedometer is a display which is common to all cars and so represents a useful starting point for discussion.

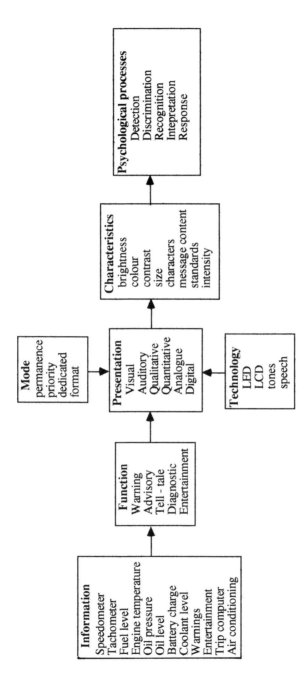

Figure 12.1   Factors defining in-car information displays.

The use and design of speedometers

Manufacturers are legally obliged to fit speedometers in their vehicles. The main reason we need a speedometer seems to be that people are often poor judges of speed when travelling in vehicles (Denton, 1966) and so require some accurate means of comparing their speed with that of statutory speed restrictions. Thus, one would expect people to use speedometers in situations which require significant alterations to be made in their driving speed. Denton (1969) compared the use made of speedometers in a number of driving conditions, such as on a motorway, exiting a motorway, driving in towns and approaching roundabouts. His results were very interesting.

1. Drivers tended to overestimate the use they made of the speedometer, believing that they referred to it far more than they actually did.
2. Drivers tended to make most use of the speedometer on motorways. It seemed that drivers set themselves a speed limit and used the speedometer to maintain this speed.
3. Drivers rarely used the speedometer when making significant changes to their driving speed, such as when approaching a roundabout. Here drivers may have been reluctant to take their eyes off the road in situations where traffic behaviour was changing rapidly.
4. When asked whether they considered a speedometer useful for driving, the majority (22 out of 30 drivers) believed that it was not.
5. Although 24 out of 30 drivers believed speed limits were important, only three drivers observed such limits while driving during the experiment.

This study raises a number of questions concerning the use of displays in vehicles. Considering that, in terms of ergonomic design, it is difficult to significantly improve on the design of the speedometer dial face, it is worth asking why do drivers not use it more often? Quite simply, the answer appears to be that, for experienced drivers at least, velocity information is derived from a range of external cues (Matthews and Cousins, 1980). Mourant and Rockwell (1972), using eye movement recordings, found a lower fixation of the speedometer among experienced than inexperienced drivers and suggested that this represents a skill acquired as part of the learning to drive process. Experienced drivers use cues from the behaviour of other traffic, their interpretation of the road conditions, e.g. the presence of periodic stripes across the road surface on approaches to roundabouts can lead to significant reductions in speed (Rutley, 1975), and the state of their vehicle. With reference to the latter point, several studies have shown that drivers are adept at using auditory information from their engine to help interpret their speed, and that removal or reduction of such information can lead to an overproduction of speed (Salvatore, 1969; McClane and Wierwille, 1975). This suggests that the speedometer needs to provide sufficient information to enable drivers to support their perceptions of their own speed, and that the information needs to be read and interpreted by drivers in as short a time as possible, in order to minimize 'eyes-off-road' time. 'Eyes-off-road' time needs to be kept short, in this instance, because drivers are receiving much of their status information from the road and traffic.

The use of digital speedometers became feasible during the 1970s and several manufacturers experimented with them. Indeed, they are still common in several American models. It is, perhaps, a little surprising that digital displays were considered, when one notes that ergonomists have been arguing for the importance of pointer position to provide additional spatial information in dial displays since the 1950s. In other words, the position of the pointer on a dial can allow the user to ascertain what range of a scale they are in, as well as indicating the number on the scale. Thus, it is not surprising to find that, while digital speedometers allow fast reading responses (Haller, 1991), analogue speedometers are more effective for decoding the speed of a car (Walter, 1991). This is because drivers tend to treat speed in relative terms, i.e. rather than attempting to maintain a precise speed, drivers will aim to keep within tolerable limits.

The notion of 'tolerable limits', in reading a speedometer, can be used to interpret some of Denton's (1969) findings. The Industrial Ergonomics Group noted that drivers will often set themselves a speed limit for a motorway and use the speedometer to maintain approximately this speed. In towns speed will be affected by the behaviour of other vehicles. If traffic appears to be moving fast drivers will also maintain a high speed. Thus, the speedometer is often redundant, with drivers preferring to use their own subjective judgements of environmental cues. It would appear, then, that the main reason for the lack of use of a speedometer relates to an over-confidence of drivers in their capabilities, especially considering Denton's (1969) findings concerning drivers' notice of speed limits. A further factor, of course, will be the relationship between attentional demands made by heavy traffic flow and a driver's desire to take his/her eyes off the road to scan a display. In heavy traffic, it might make more sense to maintain 'eyes-on-road' than on a speedometer. If this is the case, then one can appreciate the suggestion that speedometers may be redundant. However, despite Denton's (1969) research, it still unclear how drivers use their display instrumentation. How effective are the design and use of in-car warning devices in communicating potentially damaging or dangerous changes in vehicle state?

Standard warning symbology?

One might believe, from a cursory glance at car dashboards, that warning devices have achieved some sort of standardization in cars. Certainly the introduction of ISO 2575 in the 1970s was intended to reduce some of the confusion concerning symbology. Fowkes (1984) presents examples of symbols for choke control (Figure 12.2) which were used on vehicles in the late 1960s.

C. Baber

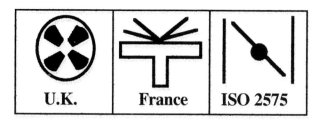

Figure 12.2   A range of symbols for 'choke'.

Simmonds (1983) argues that the use of ISO 2575 has led to a consensus of display designs among manufacturers across the world. However, this apparent consensus amongst manufacturers is governed more by the requirements of legislation than by agreement between firms or by any design principles (Baber and Wankling, 1992). Thus, even though both the International Standardization Organization and British Standards have produced recommended symbology for in-car warnings (ISO 2575; BS AU 143C), there are few guidelines concerning how the symbols should be implemented. For example, while ISO 2575 recommends the use of the colour set (red, yellow, green, blue and white) to indicate different levels of severity for warnings, advisories and tell-tales, manufacturers vary in their use of colours for different symbols or in terms of symbol placement. Furthermore, it has been noted that there are problems with the recommended symbol sets, specifically concerning interpretation of symbols and defining appropriate responses (Saunby, Farber *et al.*, 1988; Langley, 1991). Finally, it seems that legislation appears to view each symbol in isolation. Warnings are not considered as part of a system, hence the interaction between different components is not addressed (Green, 1993). This leads to the paradoxical, but often inescapable, conclusion that warnings need not provide drivers with sensible information, i.e. if drivers cannot interpret the symbols, if their interpretation does not relate to the state of the car, or if the relationship between different symbols appearing simultaneously is not clear, then the symbols cannot be said to be working effectively. The apparent lack of an ergonomic framework regarding the use and design of in-car warnings means that the addition of new symbols tends to be haphazard.

There has been a growing concern that, as the complexity of cars increases with evermore sensing devices positioned around the vehicle, information displays for drivers are becoming more and more complicated. This concern focuses upon the notion that increased display complexity may lead to a significant shift of drivers' attention from outside to inside the car (Sussman, Bishop *et al.*, 1985). In one study, Noy (1988) showed that increasing the demands of a visual secondary task (searching for short vertical lines against a field of long vertical lines) interfered with performance on a simple driving simulator. Thus, there is a serious issue concerning how best to present warning information to car drivers so as not to disrupt or otherwise interfere with their driving task. Snyder and Monty (1986) found that: '... the use of an electronic display in which the driver must timeshare his/her attention between the road and display can result in measurable reductions in lane control, speed control and display related performance'.

While the claims relating warnings to safety may be a little alarmist at present, one can foresee future technology validating such claims unless proper measures are taken. Furthermore, our concern is with the interpretation of the displays; one would assume that an in-car warning ought to signal to the driver a change in the status of the vehicle which requires attention, where failure to attend could lead to either malfunction of, or damage to, the car. Two anecdotes will illustrate these points.

The first anecdote concerns a previous generation of sophisticated in-car warnings used in an American model. A driver was cruising along a freeway when a message flashed onto the display saying 'Stop'. The driver promptly obeyed the message and was consequently sued by the drivers of the cars which piled into each other behind. The driver then successfully took the car manufacturer to court and recouped not only the monies lost from the other driver's claims, but also compensation for the incident, laying the blame upon the technology. The second anecdote concerns more conventional LED displays. Whilst driving along the motorway a driver noticed the 'battery light' come on. As there was no noticeable change in the car's performance, it was assumed that the light had simply malfunctioned. However, after stopping for petrol, the driver was unable to start the car. On opening the bonnet, it was found that the fan belt had broken; replacing the fan belt and charging the battery cured the problem.

In the first case, the driver had been presented with an explicit command to act but with no supplementary information concerning the car's status, i.e. it was not clear whether 'Stop' meant stop immediately and switch off engine, as one might expect if an oil loss suddenly occurred, or simply stop when convenient. Thus, although the designers had sought to present the information in a clear and unambiguous format, they had actually succeeded in producing a potentially confusing message. In the second case, with the benefit of hindsight and some engineering knowledge, it should be clear that the 'battery light' does not inform the driver of the status of the battery so much as the ongoing process of charging it, i.e. rather than simply concerning the battery, this light actually relates to a range of components.

It should be noted that the population who will use in-car warnings are heterogeneous, varying in age, gender, level of education and driving experience. A number of studies have found differences in performance on a task requiring interpretation of the ISO 2575 symbol set, in terms of gender, age and educational level (Green and Pew, 1978; Saunby, Farber *et al.*, 1988 and Langley, Baber *et al.*, 1992).

Efficient design of in-car warnings is not a simple matter. It should be clear that the primary aims of such warning devices should be to present relevant

Figure 12.3   Typical dashboard layout.

information in as unambiguous and informative a manner as possible so as not to interfere with the driving task. Fowkes (1984) notes that the two primary factors concerning in-car displays are visibility and legibility; it is important to ensure that the display can be both seen and read clearly. In addition there is a third, overriding factor, of intelligibility. However, it is not always clear how one should define 'relevant'. Figure 12.3 shows that, even with a small set of warnings, a typical dashboard can quickly become cluttered.

Types of 'warning' information

It is clear that not all information presented to the driver will be in the form of a warning. The role of the speedometer in a driving task has been discussed above, and it is suggested that other pointer based displays will be used in a similar fashion, i.e. intermittently and with reference to cues from the engine, such as checking the temperature indicator if an unusual smell occurs in the car, or steam from the bonnet, or checking the petrol gauge at certain points on a long journey. Grouped around these displays will be a range of lights, either simple coloured bulbs, or LEDs. The functions of the lights can be classified into three groups (Baber and Wankling, 1992), as follows:

- tell-tales – remind drivers of the status of driver controlled equipment, e.g. indicators on;
- advisories – inform drivers of future, potential problems, e.g. fuel low;
- warnings – alert drivers to immediate problems, e.g. temperature high.

While it is possible to find examples of these types of warning in all cars, it is rare to find consistent applications, as has been noted above.

The uses of in-car warning displays

It is difficult to determine the extent to which ignoring or misinterpreting in-car warning information is implicated in vehicle malfunctions. One can collect together anecdotes concerning drivers tapping the fuel gauge when it is approaching empty in order not to stop for petrol, or of speedometers getting 'stuck', but it is rare to find reports of people having trouble with the symbology on the dashboard LEDs. Either this latter finding results from people being very familiar with the LEDs, which, as has been seen above, is unlikely, or from a reluctance to admit that they do not understand the displays, or simply from the fact that people do not use these displays.

It is possible to find sound advice on designing and positioning symbols which are conspicuous and legible (Easterby and Zwaga, 1984), and many in-car warning symbols can be seen to follow such advice. It is a more difficult task to determine how the symbols are used – what is the relationship between level of engineering knowledge and symbol interpretation; if a driver can deduce a fault from external cues, will a symbol be necessary, and if the driver cannot perform such diagnosis, will the symbol be meaningful?

It is difficult to determine the reasons for adopting warning symbology in cars. Our main sources seem to suggest that symbols were used because they could fit onto the space made available by the replacement of bulbs with LEDs and that, in some cases, such as on stalks, symbols would be the only form of information presentation which could sensibly be used. Thus, the prime motive appears to be related to available space. Later justifications concerned the potential variation in languages of car buyers, with symbols proposed as an international language and easy to interpret. However, some of the problems people have in using these symbols have been noted, and one suggestion is that they could either be redesigned to take account of the information needs of drivers, or replaced. If symbols were to be replaced, therefore, the new formats would need to take note of limited space requirements and provide multi-language capabilities.

Novel formats for in-car warning displays

In order for in-car warning displays to become more ergonomic, it is necessary to reconsider the philosophy underlying their design. In this section, we briefly review our own work on reconfigurable displays, before discussing forms of auditory warning.

Reconfigurable visual displays

The basic idea of reconfigurable displays is that all information will be presented on a single display space, with information appearing on a single screen. A number of production cars already employ such technology and several manufacturers are experimenting with alternative designs. We have decided to step back from technological and application issues and consider the information requirements of such displays. We have noted that people have difficulty in interpreting many of the standard symbols (ISO 2575) used in cars (Langley, 1991). Furthermore, we have found that, ironically, it is the least important symbols which are most commonly recognized, and the most important which people have difficulty with (Langley, Baber *et al.*, 1992). This is probably a function of familiarity, on the one hand, and the inherent 'concreteness' of the symbol, on the other, with the more pictorial symbols being easy to recognize. For example, the least important symbols are often the ones which occur most often, e.g. indicators and, as an example of concreteness, 'fuel low' is indicated by a petrol pump. The less concrete symbols can be relatively easy to interpret, given minimal engineering knowledge, e.g. the ISO 2575 symbol for brakes looks like a drum brake. Thus, it might be possible to propose a short training session for people to become familiar with the warning symbols when they buy a new car. However, this proposal is clearly impracticable, especially when one considers the huge trade in second-hand cars. An alternative to training would be to use expanded display capabilities to enhance the symbols.

In one study, combinations of symbols and supplementary text were used to assess interpretation of the urgency of warnings. This study is reported in detail in Baber and Wankling (1992). Our main recommendations were simply that performance improved if symbols were supplemented by text informing users of the appropriate course of action required. A further study examined the use of colour coding in warning symbols. It was found that a colour set of red for warning, amber for advisory and green for tell-tale was easy for subjects to learn and facilitated performance (Goodsman, 1991). However, interestingly enough, the provision of colour did not enhance interpretation of the content of the symbol, only its perceived urgency. As there does not appear to be an agreed convention among car manufacturers concerning the colours used on LEDs, this finding suggests that some standardization is required.

In both of these studies, there was a strong implication that the symbols themselves conveyed little or no useful information. This notion was tested in a paper based experiment, in which performance using a combination of symbol plus text, or symbol alone, or text alone was assessed. Forty subjects completed the exercise, with 13 subjects in the text and symbol only groups, and 14 in the combination group. While both text only and combination performance was better than symbol alone, there was no difference between the former two groups. This suggests that, in these studies, the presence of the symbol was, at best, redundant. While it could be objected that the use of language based displays will limit the potential market, judicious use of on-line programming at the installation stage could create displays for specific countries. Obviously this would require some careful consideration. An investigation would be required into what would be meaningful warning terms to provide necessary and sufficient information to drivers. One could also suggest that text based displays might be more demanding of drivers attention, because they contain more information. Although Baber and Wankling (1992) did not find this to be the case, further work is planned in which these displays' configurations are examined in performance of driving tasks, using a driving simulator.

Auditory displays

There are many situations in which a driver will use auditory information from the engine, e.g. when changing gear, or to diagnose engine malfunction, or from other parts of the vehicle – as when noting that a door is not closed properly – or concerning the status of turn indicators or hazard lights. There has been some suggestion that the provision of auditory warnings can also reduce the visual load of the driver (Wolf, 1987). In her study, Wolf (1987) defined three main types of auditory display and gave a brief description of their characteristics. A turn signal would be a simple, short sound with regular pattern and even tempo. A reminder sound would consist of a simple musical chime with one or two tones. A warning sound would be a short, sharp chime, which increases in volume as the problem becomes more serious. Naturally there is a need to determine the correct intensity of such signals in the, often noisy, car environment. Antin, Lauretta *et al.* (1991) have found that the preferred level of intensity is generally higher than the level required for detection. They also found that, with the presence of noise from a radio, there was an increase in false alarms and a high variability in performance. One potential problem of auditory warnings could be their propensity to startle the driver at inappropriate moments, which could lead to accidents. In this instance, the strength of the warning type could also be its weakness. However, one cannot rule out the possibility that auditory warnings could be designed to either overcome or reduce this problem.

Speech displays

There has been much media attention devoted to the 'speaking car'. For instance, the British Leyland 'Maestro' had a simple voice synthesis system which could announce low oil pressure, high engine temperature or door open (Redpath, 1984). There were, however, a number of complaints regarding this system, which led to some drivers disabling the speech system. We would suggest that this application failed on four counts. Firstly, it did not distinguish between different levels of advisory and warning messages. Thus, there may have been a tendency either to over-emphasize unimportant information, or to ignore important information as the result of the system 'crying wolf'. Secondly, if there was a malfunction in one of the sensors, e.g. if its tolerance was set too low so that it kept tripping, the speech display would continually send inappropriate messages. For visual displays, this need not present a problem as the driver can potentially habituate to it, but it can be difficult to habituate to the rough tones of a speech synthesizer. Thirdly, the notion that a speech display would not intrude upon the primary task of driving, which seems the main foundation for this application, has not been empirically verified nor is it intuitively plausible – speech syntheses are very effective at capturing attention immediately and, like other auditory displays, could startle the driver. Fourthly, the quality of the synthesis used in the Maestro simply was not very good. As with the use of auditory displays, it is difficult to determine the effects of radio noise on the perception of a speech display.

Conclusions

To a large extent, the points raised in this paper are not specific to cars but have been discussed with reference to other topics in this book. However, it is worth noting the particular characteristics of in-car warnings in the driving task – they create a situation in which people are presented with copious quantities of information which they do not want and which, often, they do not know how to use, coupled with the highly skilled activity of driving a ton of metal at speed. The issue is less one of defining the appropriate format for displays and more one of finding the appropriate information which needs to be displayed, and the timing of such display.

References

Antin, J.F., Lauretta, D.J. and Wolf, L.D., 1991, The intensity of auditory warning tones in the automobile environment: detection and preference evaluations, *Applied Ergonomics*, **22**, 13–19.

Baber, C. and Wankling, J., 1992, Human factors considerations of reconfigurable displays for automobiles: an experiment into the use of text and symbols for in-car warning displays, *Applied Ergonomics*, **23**, 255–62.

Denton, D.G., 1966, A subjective scale of speed when driving a motor vehicle, *Ergonomics*, **9**, 203–10.

Denton, G.G., 1969, The use made of the speedometer as an aid to driving, *Ergonomics*, **12**, 447–54.

Easterby, R. and Zwaga, H., 1984, *Information Design*, Chichester: Wiley.

Fowkes, M., 1984, Presenting information to the driver, *Displays*, **5**, 215–23.

Goodsman, I., 1991, The Use of Colour in Reconfigurable Displays, Birmingham University School of Manufacturing and Mechanical Engineering, Ergonomics Project Report.

Green, P., 1993, Design and evaluation of symbols for automobile controls and displays, in Peacock, B. and Karwowski, W. (Eds) *Automotive Ergonomics*, London: Taylor & Francis, pp. 237–268.

Green, P. and Pew, R.W., 1978, Evaluating pictographic symbols: an automotive application, *Human Factors*, **20**, 103–14.

Haller, R., 1991, Experimental investigation of display reading tasks in vehicles and consequences for instrument panel design, in Gale, A.G., Brown, I.D., Haslegrave, C.M., Moorhead, I. and Taylor, S. (Eds) *Vision in Vehicles III*, pp. 197–211, Amsterdam: North Holland.

Knoll, P.M., Heintz, F. and Neidhard, K., 1987, Application of graphic displays in automobiles, *Society for Information Display Symposium of Technical Papers*, Vol. XVIII, pp. 41–4, New York: Palisades Institute for Research Services.

Langley, D., 1991, 'Investigations into In-Car Reconfigurable Warning and Tell-Tale Displays, unpublished MSc thesis Birmingham University School of Manufacturing and Mechanical Engineering.

Langley, D., Baber, C. and Wankling, J., 1992, Reconfigurable displays for in-car information systems, in Lovesey, E.J. (Ed.) *Contemporary Ergonomics 1992*, pp. 29–34, London: Taylor and Francis.

Matthews, M.L. and Cousins, L.R., 1980, The influence of vehicle type on the estimation of velocity while driving, *Ergonomics*, **23**, 1151–60.

McClane, R.C. and Wierwille, W.W., 1975, The influence of motion and audio cues on driver performance in an automobile simulator, *Human Factors*, **17**, 488–501.

Mourant, R.R. and Rockwell, T.H., 1972, Strategies of visual search by novice and experienced drivers, *Human Factors*, **14**, 325–35.

Noy, Y.I., 1988, Selective attention and performance while driving with intelligent automobile displays, in Adams, A.S., Hall, A.R., McPhee, B.J. and Oxenburgh, M.S. (Eds) *Proceedings of the 10th. Congress of the International Ergonomics Association*, pp. 587–9, London: Taylor and Francis.

Redpath, D., 1984, Specific applications of speech synthesis, *Proceedings of the First International Conference on Speech Technology*, Bedford: IFS.

Rutley, K.S., 1975, Control of drivers' speed by means other than enforcement, *Ergonomics*, **18**, 89–100.

Salvatore, S., 1969, Velocity sensing – comparison of field and laboratory methods, *Highway Research Record*, **292**, 79–91.

Saunby, C.S., Farber, E.I. and deMello, J., 1988, *Driver Understanding and Recognition of Automotive ISO Symbols*, SAE Technical Paper Series paper number 880056, Warrendale, PA.

Simmonds, G.R.W., 1983, Ergonomics standards and research for cars, *Applied Ergonomics*, **14** (2), 97–101.

Snyder, H.L. and Monty, R.W., 1986, A methodology for road evaluation of automobile displays, in Gale, A.G., Freeman, M.H., Haslegrave, C.M., Smith, P. and Taylor, S.P. (Eds) *Vision in Vehicles*, Amsterdam: North Holland.

Sussman, E.D., Bishop, H., Maknick, B. and Walter, R., 1985, Driver inattention and highway safety, *Transportation Research Record*, 1047.

Train, M.H., 1987, Advanced instrumentation of the Oldsmobile Aerotech, *Society for Information Display Symposium of Technical Papers*, Vol. XVIII, pp. 37–40, New York: Palisades Institute for Research Services.

Walter, W., 1991, Ergonomics information evaluation of analogue versus digital coding of instruments in vehicles, in Gale, A.G., Brown, I.D., Haslegrave, C.M., Moorhead, I. and Taylor, S.P. (Eds) *Vision in Vehicles III*, Amsterdam: North Holland.

Wolf, L.D., 1987, The investigation of auditory displays for automotive applications, *Society for Information Display Symposium of Technical Papers* Vol. XVIII, pp. 49–51, New York: Palisades Institute for Research Services.

Zwahlen, H.T. and deBald, D.P., 1986, Safety aspects of sophisticated in-vehicle information displays and controls, *Proceedings of the 30th. Annual Meeting of the Human Factors Society*, pp. 256–60, Santa Monica, CA: Human Factors Society.

Sources of confusion in intensive therapy unit alarms

Christina Meredith and Judy Edworthy

Introduction

The intensive therapy unit has been likened to a war bunker (Hay and Oken, 1972). There are periods of clam and inactivity, interspersed with periods of heavy and demanding work. To aid nurses and doctors during both periods of activity a great deal of sophisticated technology is used. Almost all pieces of equipment are armed with displays and alarms in an attempt to help reduce the workload. As verbal indicators and warnings are inappropriate in this environment, non-verbal auditory warnings are used a great deal. However, the use of alarms is generally both excessive and inappropriate, and attempts are being made to redress the problem amongst standard-generating organizations such as the British Standards Institution (BSI); the Comité Européen de Normalisation (CEN), and the International Stand-ardization Organization (ISO). In this chapter a project currently in progress at the University of Plymouth will be presented, which shows potential sources of auditory confusion. This has important implications for future alarm de-sign and standardization.

The documented problems associated with the use of auditory warnings in the hospital environment are not substantially different from those found in other high-workload environments. Like the cockpit of a helicopter, or the control room of a nuclear power plant, there are usually too many warnings; they are often too loud (or conversely, inaudible); they are confusing, and there is often no consistent rationale associating the sounds with their mean-ings. In the ITU, intensive therapy unit, a critically ill patient may be con-nected to several pieces of equipment. Each piece of equipment has its own alarm sound and may have different noises depending on what the problem is. Furthermore, each piece of equipment may be produced by several different manufacturers, each manufacturer incorporating its own auditory alarm into the monitoring equipment. Exceeding 20 or so alarms per patient is easily done, and with this number confusion can arise even with a sensible and ergonomic rationale for alarm design and implementation. As proper design and implementation is often found wanting, additional sources of confusion come from the warning sounds themselves. Manufacturers tend to use their own preferred alarms and these may not integrate with others. A situation can easily be envisaged where alarms are associated by manufacturer and not by function; similar-sounding alarms could come from equipment with totally different functions because they are made by the same manufacturer. Equipment with related functions, but from different manufacturers, could have completely different sounds.

Many of the problems associated with auditory warnings in hospital environments are psychological in nature. A study by Patterson and Milroy (1980) showed that subjects find it easy to learn five or six warnings, beyond which committing new warnings to memory becomes much more difficult. Also, warnings are often confusing. They are confusing not only because there are too many of them, but also because many warnings have continuous, high-pitched tones. As pitch judgement tends to be a relative, rather than an absolute, judgement and information is lost very quickly about the absolute values of pitches (Deutsch, 1978), then the vast majority of hospital staff who do not possess absolute pitch are at a disadvantage in discriminating warnings of this sort. Moreover, warnings are often inappropriate in terms of their 'urgency mapping' (Momtahan and Tansley, 1989). There is generally no relationship between the urgency of a medical situation and the perceived urgency of the alarm which signals that condition. The psycho-acoustic urgency of a warning might not be important if the meaning of the warning is already known. However, in many instances it is not known (Momtahan, Tansley *et al.*, 1993). Let us illustrate this with an example. In the intensive therapy unit at Derriford Hospital, Plymouth, one of the food pumps used has an extremely loud, urgent-sounding alarm. In contrast, the alarms of the ventilators are quieter, less urgent-sounding alarms. However, in terms of importance, the ventilator is more essential in maintaining the life support of the patient than the food pump. While experienced staff will have learned which sound is more important, new staff will not and may assume that the food pump, because it sounds so urgent, is an urgent problem that needs attending to immediately. Urgency mapping is considered in detail in a separate chapter (Edworthy, this volume).

From a practical point of view it is important to know which sorts of warnings are confused, because deaths have been attributed to confusions between pieces of equipment with similar warnings (Cooper and Couvillon, 1983). However, it is essential not only to know how confusions occur when subjects are learning warnings in a laboratory experiment, when they are doing nothing except attending to the learning task, but also it is important to know how confusions occur when nurses are working, perhaps under stress, in the intensive therapy ward when they may be attending to a number of other tasks. In order to do this, has to be found some way of measuring and evaluating their work demands in order for any study to be truly ecologically valid and applicable. This paper describes the first part of a project currently under way at the University of Plymouth in which the relationship between auditory confusion and workload is being evaluated. The eventual aims of the project are to carry out a study of auditory confusion in the intensive therapy ward, and to make recommendations for the future design and implementation of warnings. It is predicted that, as workload increases, so will auditory confusion. It is possible also that the pattern of confusion will alter as workload increases. This chapter however, reports only on the laboratory-based experiments of auditory confusion as the workload study is currently in progress.

Alarm confusion

The investigation of confusions between auditory warnings would seem to be a laboratory-based task from which principles about confusion could be generalized to the working environment. However, caution is needed even here if such generalizations are to be made. For example, a study by Patterson and Milroy (1980) required subjects to learn a set of aircraft warnings in both a cumulative and a paired-associate learning paradigm. Results showed that the learning rate slowed significantly after five or six warnings had been committed to memory; the study also showed that warnings were most often confused when they shared the same repetition rate or temporal pattern, regardless of other acoustic differences. However, experimentation on the learning and retention of warnings may be compromised by the techniques of experimental psychology themselves. Memory for warnings may be better than experiments might suggest, but the methods of experimental psychology dictate that information which may be of use in the real environment is eliminated in the laboratory. For example, in a laboratory-based study, the stimuli would all be equalized for loudness, length and so on in order to isolate properly those acoustic qualities which cause confusion. Whilst it is perfectly right to control for such features in the name of good experimental psychology, it is important to consider how people might discriminate between auditory warnings in practice and to allow these cues into the laboratory. For example, it is possible that the most important cues used by people in differentiating between warnings might be differences in loudness and length, with more subtle acoustic cues perhaps being only of secondary importance.

Table 13.1   Stimuli

Sounds	Names Exp. 1 and Exp. 2	Names Exp. 3	Equipment
1	Aircall	Alpha	Erica ventilator
2	Bleep	Bravo	Doctor's bleep
3	Heart rate	Delta	Electrocardiogram
4	Fire	Echo	Doctor's 'crash' bleep
5	Humidifier	Foxtrot	Fisher-Pago humidifier
6	Infusion	Golf	3M 200 infusion pump
7	Oxymeter	Hotel	Simed pulse oxymeter
8	Perfusion	Kilo	Brompton ventilator (2)
9	Pulse	Lima	Pulse oxymeter 505
10	Syringe	Sierra	Syringe driver
11	Ventilator	Tango	Brompton ventilator (1)
12	Respiratory	Uniform	Vickers syringe pump

Experiments on auditory confusion

Introduction and method

In order to elicit a clearer picture of auditory confusion, three experiments were conducted in which the stimuli and procedures were almost identical except for small, but important, differences. A set of 12 auditory warnings currently in use in the intensive therapy unit at Derriford Hospital, Plymouth, were recorded on DAT (Digital Audio Tape) and transferred onto computer prior to experimentation. In the first experiment the sounds were retained in exactly the same form in which they had been recorded from the hospital ward. Consequently in this set some warnings were louder than others, some were longer, and some repeated whereas others did not. This set of warnings is referred to as the 'natural' warnings. The second set of stimuli were standardized so that they were all of the same length and the same loudness in a way typical of experimental psychology methods. This set of warnings is referred to as the 'controlled' warnings. The third experiment used the same sounds as those used in the second experiment but the names were changed. This was to ensure that any confusions made were acoustic and not semantic. These sounds are referred to as 'neutral names'. In experiment 3 the names were re-ordered halfway through the experiment to ensure that no order effect had occurred. The warnings are described in Table 13.1 and Table 13.2. Table 13.1 shows the equipment with which the warnings are normally associated and the names used for these warnings in the three experiments. Table 13.2 gives a brief acoustic description of each warning.

The stimuli were then tested in three experiments, for which the procedure was identical. In each experiment, subjects participated in two experimental sessions. In the first, they learned the 12 warnings in a paired-associate paradigm. Each warning was first presented to the subject and named. When all 12 warnings had been presented, subjects heard each one in turn in random

Table 13.2   Description of stimuli

Sound	Description of sound
1	Two-tone pattern: first tone = 0.5 secs at 500 Hz; second tone = 1.0 secs at 250 Hz (including a short gap of 0.2 secs)
2	Total sound lasts for 5 seconds; tone 0.7 secs at 280 Hz followed by a repeating two-tone pattern at 2600 Hz and 2000 Hz. Sound then repeats after a gap of 7 secs
3	1000 Hz continuous tone
4	Tone of 1 second at 260 Hz, followed by burst of short tones each lasting 0.25 secs at 2600 Hz
5	Repeated 4 pulse pattern = 0.6 secs: first two pulses at 1000 Hz; second two pulses at 930 Hz; each individual pulse = 0.15 secs
6	'Beethoven's 5th': 3 pulses at 280 Hz + 1 at 220 Hz; each pulse = 0.15 secs; total length = 0.6 secs + 0.6 sec gap
7	Two pulses lasting 0.5 secs at 260 Hz and 275 Hz, with a pause between pulses of the same time interval (0.5 secs)
8	Repeated tone lasting 0.5 secs at 1760 Hz; 1 second pause between tones
9	Two-tone repeating pattern: first tone = 0.25 secs at 400 Hz; second tone = 0.25 secs at 310 Hz
10	Complex pattern of long tones (1 second) and very short pulses (100 ms) all at 3000 Hz
11	Repeated tone lasting 0.35 secs at 2200 Hz; pause of 0.35 secs between each tone
12	Continuous tone at 2350 Hz with strident higher harmonics

order and were required to name the warning. If named incorrectly, correct feedback was given immediately. This procedure was repeated 12 times, so that subjects were required to name each warning on 12 occasions. A week later, subjects were retested in a return phase. The warnings were presented one at a time, again in random order, in sets of 12, until all warnings were correctly re-identified.

Results

Three main features of the data are of interest: firstly, the number of trials taken to identify correctly all of the warnings; secondly, the total number of times each warning was identified correctly; and thirdly, the pattern of errors (i.e. the response given when a warning was incorrectly identified). Figure 13.1 shows the number of trials taken by each subject to learn the warnings in the learning and return phases of the three experiments.

Participants took significantly fewer trials to reach criterion in experiment 1, in comparison to both experiments 2 and 3. Experiment 2 and experiment 3 did not differ significantly.

The results therefore show that participants found it easiest to learn the sounds in the 'natural' warning experiment (experiment 1) and most difficult

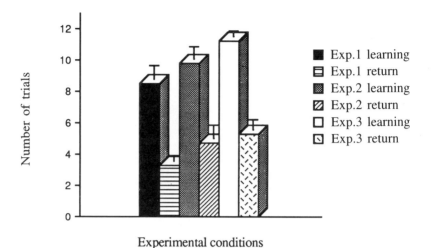

Figure 13.1   Number of trials taken to identify correctly all twelve warnings, experiments 1, 2 and 3.

in the 'neutral names' experiment (experiment 3) where many of the cues had been removed, including the usual names. It appears that subjects took longer to develop a mnemonic strategy in the final experiment. For example, a large proportion of subjects in experiment 1 and experiment 2 remembered sound '3' (heart monitor) by imagining a sound commonly heard on television dramas where a person attached to a monitor has died, and a long continuous tone is heard. In the 'neutral name' condition, the name 'delta' was used for this sound and no specific mnemonic strategy was used generally by the subjects.

The differences between the sounds becomes apparent when the second main feature of the data is considered. Taken as a mean of the total number of possible correct responses, the means of the correct responses to the 'natural' sounds (experiment 1) was significantly greater than that for the 'controlled' sounds (experiment 2) in both the learning and the return phase. This is shown in Figures 13.2 and 13.3. Figures 13.4 and 13.5 show the learning and return phases for experiment 2 (controlled sounds) and experiment 3 (neutral names). The considerable differences between experiment 1 (natural sounds) and experiment 3 (neutral names) are shown in Figures 13.6 and 13.7.

Statistical comparisons were made between experiment 1 (natural sounds) and experiment 2 (controlled sounds) as the names of the sounds were the same (see Table 13.1) and between experiment 2 (controlled sounds) and experiment 3 (neutral names) as the sounds were the same (i.e. they were matched for length and loudness). Although interesting, comparisons between experiment 1 (natural sounds) and experiment 3 (neutral names) are

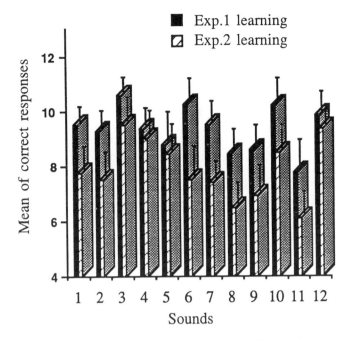

Figure 13.2   Experiment 1 and experiment 2 (learning phase).

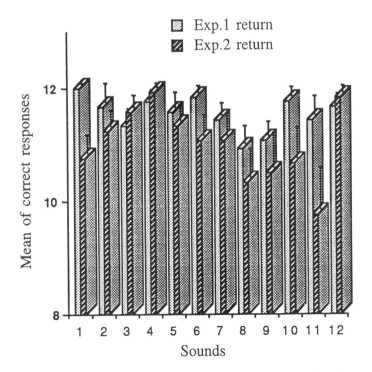

Figure 13.3   Experiment 1 and experiment 2 (return phase).

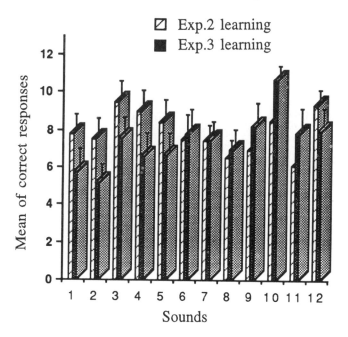

Figure 13.4 Experiment 2 and experiment 3 (learning phase).

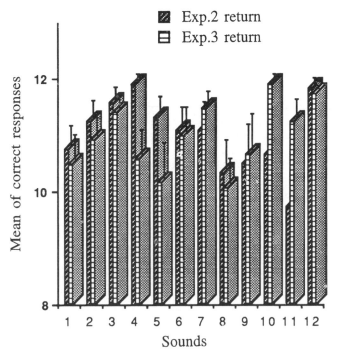

Figure 13.5 Experiment 2 and experiment 3 (return phase).

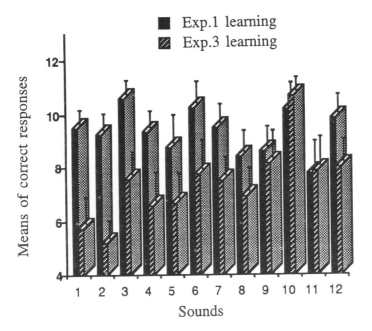

Figure 13.6   Experiment 1 and experiment 3 (learning phase).

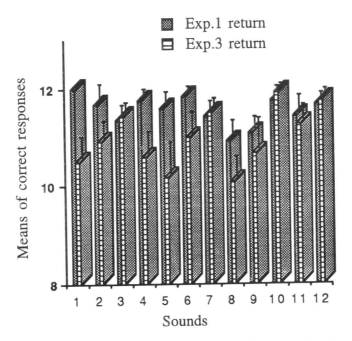

Figure 13.7   Experiment 1 and experiment 3 (return phase).

Table 13.3 Significant confusions (p < 0.01) between warnings

Sounds played	Sounds identified as:					
	Exp 1 Learning	Exp 1 Return	Exp 2 Learning	Exp 2 Return	Exp 3 Learning	Exp 3 Return
1	7, 11			2		
2		5		1	4, 10	
3	12	12	12	12	6, 12	
4	2, 8		2		2	
5		9	9, 10			9
6	5		1, 7, 10			
7	9	6	6	9		1
8	11	7, 11	5, 10, 11, 12		1, 9, 11	11
9		5	12	7		5
10				5		
11	8	8	9	2	1, 8, 12	8
12	3, 11		3, 11		3	

not legitimate for statistical purposes as they differ in two important ways: names of the sounds are different (neutral names and original names); and the sounds are different (natural and controlled).

Two 3-way analyses of variance (Experiment (two levels) × Phase (two levels) × Sound (12 levels)) showed that the mean number of correct responses was significantly different between experiments 1 and 2, but not between experiments 2 and 3. The mean number of correct responses was significantly higher in the return phase in all experiments and there were significant differences between individual sounds, again in all experiments.

Thus there appears to be some learning and retention advantage associated with the natural sounds. This is not altogether surprising, because subjects have all the cues associated with the controlled sounds as well as some additional ones. However, it is important to demonstrate this experimentally, because cues of this sort are usually left out of experimental studies. The results suggest that warnings are more easily learnt when heard in their naturally occurring, rather than their experimental, form. This is useful to know for practical purposes.

The other main focus of interest in these experiments are the confusions between warnings. Using a method of analysis based on the multinomial distribution, which also takes into account response bias, significant confusions were isolated. The significant confusions between warnings in both the learning and return phase of all three experiments are shown in Table 13.3. The number of highly significant confusions ($p < 0.01$) are relatively small, and these are shown in the table. The number of confusions at a lower level of significance ($p < 0.05$) is considerably higher, especially for the learning phase of experiment 1 (Meredith, 1992). The pattern of confusion between experiments is also slightly different.

Discussion

One point of interest is that most of the confusions are asymmetric, for example sound '4' (Fire) was frequently named as sound '2' (Bleep) but not vice versa ('2' was not named as '4' in the same way). The only clear cases where confusion was symmetrical in the learning phase were between warnings '3' and '12' (all experiments) and '8' and '11' (symmetrical for experiment 1 and experiment 3 but not for experiment 2). The acoustic similarities between these pairs of warnings shows why these warnings are confused. Warnings '3' and '12' are both continuous tones, and they are the most readily confused of all of the warnings. It is interesting to note that although they are approximately an octave and a third apart in pitch, and when heard together can be clearly discriminated, subjects confuse them when they are heard with longer time intervals between them (the time intervals would be much greater in practice than in the experiment). This confusion shows that subjects find it difficult to discriminate between warnings on the basis of pitch alone if they share other characteristics such as temporal pattern (in this case, a continuous sound with no temporal characteristics).

The confusion between warnings '8' and '11' is also of some interest. Both of these warnings have a regular on/off temporal pattern, so in that sense they are similar. However, warning '8' has a much slower pulse speed than warning '11', with the on/off cycle of warning '8' being about 2.5 times slower than warning '11'. Even with this large difference, the two warnings are consistently confused.

These two confusions are the most striking of the confusions found across all three experiments. However, there are many others which are discussed elsewhere (Meredith, 1992). Of particular interest are confusions between warnings which begin with a long tone and continue in a completely different fashion (for example, warnings '2', '4' and '8'), and those which have a succession of very short pulses but differ substantially in other ways (warnings '10', '5' and '6').

Some of these results suggest that the basis on which confusions can occur might be related to the sorts of labels subjects use to describe, and memorize, the warnings rather than to their precise acoustic qualities. For example, some of the confusions we have found are more readily understood if we assume that subjects used the label 'high-pitched, continuous tone' to memorize warnings '3' and '12', and 'regular on/off' for warnings '8' and '11'. For warnings which are acoustically quite different, the simple label 'complex' or 'musical' might have caused confusion. This issue requires further detailed investigation but is not currently of major concern in the research project.

In summary, our results show that natural sounds are more often identified correctly than those which are 'sanitized' for the purposes of experimental work. This may explain why Patterson and Milroy (1980) found that subjects could only learn five or six warnings. The results also show that subjects are in fact quite good at learning and remembering warnings, with relatively few confusions, even up to 12 warnings. All subjects learned a set of 12 warnings within one hour, many in a significantly shorter time. This has a very clear and simple message for working practice: a little time spent teaching warnings to medical staff would pay dividends.

On the topic of auditory confusion itself, our results suggest that confusion can occur even between warnings that are quite different at the level of acoustic description (Table 13.1). This may be attributable to the labels that subjects inevitably use in memorizing the sounds during the learning phase. The more enduring confusions (Table 13.3, return phases) are still of a rather gross nature, and not perhaps as subtle as earlier work (Patterson and Milroy, 1980) would suggest.

Design implications

One use of this work could be to develop guidelines and proposals for a set of auditory warnings for intensive care work which would be less confusing than those currently available; this may be possible in the future through standardization work. However, for the purposes of our project we simply need to isolate the significant confusions between the warnings when learned in the laboratory under ideal conditions, in order to provide a baseline for comparison when warnings are learnt and heard under conditions in which other tasks are being performed at the same time. Generally, standardization of warning sounds now includes some concept of urgency mapping but still remains unclear on the issue of confusion and discrimination; statements to the effect that warnings sets should be as discriminatory as possible abound but standards rarely, if ever, give guidance on how to achieve this. Our work will also contribute to the improvement of warnings discrimination and the specification of warning sets.

Acknowledgements

This research was supported by a SERC studentship. The authors wish to thank ITU Staff, Derisford Hospital, Plymouth for their assistance.

References

Cooper, J.B. and Couvillon, L.A., 1983, 'Accidental breathing system disconnections,' interim report to the Food and Drug Administration, Cambridge.

Deutsch, D., 1978, Delayed pitch comparisons and the principle of proximity, *Perception and Psychophysics*, **23**, 227–30.

Edworthy, J., Urgency mapping in auditory warning signals, this book.

Hay, D. and Oken, D., 1972, The psychological stresses of intensive care nursing, *Psychosom Med.*, **34**, 109–18.

Meredith, C.S., 1992, 'Auditory confusion in the ITU', unpublished technical report, University of Plymouth.

Momtahan, K.C. and Tansley, B.W., 1989, 'An ergonomic analysis of the auditory alarm signals in the operating room and recovery room', presentation at the Annual Meeting of the Canadian Acoustical Association, Halifax, Nova Scotia, October.

Momtahan, K.C., Tansley, B.W. and Hetu, R., 1993, Audibility and identification of auditory alarms in operating rooms and an intensive care unit, *Ergonomics* (in press).

Patterson, R.D. and Milroy, R., 1980, Auditory warnings on civil aircraft: the learning and retention of warnings, *Civil Aviation Authority report number 7D/S/0142.*

14

Key topics in alarm design

Neville Stanton

Introduction

This book has considered human factors issues relating to industrial alarm systems from four viewpoints: laboratory investigation, the existing literature, current practice and prospective developments. From the preceding chapters, nine key topics have emerged, namely: legislation, types of alarm systems, problems with alarm systems, alarm reduction, human factors approach, definitions, human supervisory control, alarm initiated activities and characteristics of alarm media. These key topics form the bases for the discussions in this chapter.

Legislation

In chapter 1 it was indicated that designers of alarm systems face legislative requirements to consider the human factors issues in design. The outcome is that where existing or proposed systems are inadequate, human factors will need to be employed. This has two principal effects. Firstly, it raises the profile of human factors in the design of alarm systems. Secondly, it forces engineers to recognize that human factors is necessary and integral to the design process.

This legislation presents a great challenge to human factors to show what it has to offer the designer and engineer. Failure to live up to the promise could have a negative effect on the discipline. However, if this potential is realized, the discipline will undoubtedly spread its influence throughout the engineering and computing disciplines. At present standards are being developed, such as those on software ergonomics and the man–machine interface. Standards like these offer the designer some guidance and, inevitably, are built upon a foundation of human factors.

Types of alarm systems

This book has suggested that although legislation demands that alarms are unambiguous, easily perceived and easily understood, many industrial systems do not live up to this expectation. Examples of alarm systems falling short of ideal are provided in most of the chapters. The majority of the chapters present discussions of alarms systems within human supervisory control tasks, i.e. operation of plant via a remote control room in power generation, oil and chemical production and manufacturing. However, other domains are mentioned, such as aviation, automobiles and medical units.

Consideration of the different alarm systems presented within the book demonstrate that there are some common problems despite the quite different application areas. Typically, most of these problems relate to the way in which the alarm media are used and the type of information displayed. There tends to be a basic conflict between the manner in which information is displayed and the goal the person using the information is seeking to achieve.

Problems with alarm systems

Problems associated with industrial alarm systems suggest that the medium is used inappropriately. Edworthy (chapter 2) argues that the problems stem from a 'better safe than sorry' philosophy of alarm design. This approach leads designers to design too many alarms that are too loud and too insistent. This can make auditory warnings confusing, due to similarity and inappropriate urgency mappings (Meredith and Edworthy, chapter 13). Baber (chapter 12) suggests that visual warnings can also be confusing and proposes that standardization of warning symbology is desirable.

Hollywell and Marshall (chapter 3) point out that the rapid onset of a large number of alarms can make diagnosis difficult. They indicate that high alarm arrival rates, in the region of 100 alarms per minute for the duration of three minutes can occur. This needs to be considered in the light of a plant that may have up to 18 000 alarms (Marshall and Baker, chapter 11). The sheer amount of information presented to the operators can make the task of tracking the progress of an incident difficult (Zwaga and Hoonhout, chapter 7). This 'cognitive overload' associated with heavy transients is, in part, due to:

• repetitive alarms;
• precursor alarms;
• standing alarms;
• consequence alarms;
• oscillatory alarms.
  (Bye, Berg *et al.*, chapter 9)

Hickling (chapter 10) argues that designers need to consider the operational needs of the operators if they are to design alarms systems that are to be successful in human factors terms.

Alarm reduction

One might consider that an appropriate solution to many of the problems associated with the alarm systems could be dealt with through alarm reduction techniques. This would seem to make the genuine alarms more obvious and reduce the masking phenomenon. Bye, Berg *et al.* (chapter 9) indicate the type of methods that could be employed to rid the alarm display of the repetitive, precursor, standing and consequence alarms. They also mention a logical alarm reduction system that has been developed called HALO (Handling Alarms using LOgic – this system was also introduced by Marshall and Baker in chapter 11). Bye, Berg *et al.* argue that logic filtering techniques can be an effective means of reducing the number of active alarms during transients. Hickling (chapter 10) also proposes logical suppression techniques as offering a promising means of reducing 'cognitive overload' of operators during heavy transients. However, Zwaga and Hoonhout (chapter 7) caution the introduction of alarm suppression methods without first evaluating and validating the effects to ensure that the system still meets the operational needs of the operators. Further, Marshall and Baker (chapter 11) argue that apparently redundant alarms can act as a source of reassurance to the operator. The operator uses the alarm system to check that the plant is behaving as expected: in the words of Marshall and Baker 'the alarm may be anticipated but it is not a nuisance'.

Human factors approach

It is suggested that the absolute number of alarms presented is not the central issue to alarm design from a human factors perspective. A human factors approach would consider the operator's ability to manage the situation safely and efficiently as the most salient question. This brings a fresh approach to the design of alarm systems. Human factors has its own perspective and methods. It also draws on an expansive body of knowledge. This multi-faceted approach allows human factors specialists to get to grips with a problem on many fronts, to have a more complete understanding of the problem and to get an insight into the means of dealing with it.

The introduction of information technology into the control room has not always gone hand-in-hand with improved task performance. This is not because it is unsuitable, rather it is often due to the maximization of information to the operator without accommodating for the limits, or exploiting the

Table 14.1   Dimension of alarm assessment methods

Focus of assessment	Nature of assessment	
	Static	Dynamic
Operator	Questionnaires	Scenarios
Technology	Static assessments	Observations

potential, of human performance. In addressing the question of what to alarm, one should consider to whom the information would be useful. Alarms that are of use to the maintenance engineer are not necessarily going to be useful to the plant operator.

Typically, these are mixed in the same system, providing the plant operator with a lot of irrelevant information that could potentially mask more important alarms. Similarly, defining thresholds to trigger alarms requires careful fine tuning. Unfortunately, plant commissioning is often a hurried process, leaving the operator with many 'false' alarms that can be attributed to design failures. Presentation of the information may be largely dictated by screen capability and hardware capacity, rather than by human performance. However, the optimum method of presenting information will be determined by what the operator will be required to do with it. This concern goes beyond the initial detection of faults, to consider the use operators make of the alarm system, particularly when dealing with perturbations and incidents. Therefore, it is worthwhile considering methods of evaluating the alarm system and the use that the operator makes of it. There are many approaches that can be used to evaluate an alarm system. These include the use of questionnaires to examine operator reactions to their alarm systems; observation forms for the quantification of the alarm messages generated by the system, static assessments of alarm panels and talk-through scenarios. They differ in terms of the nature of the assessment (static or dynamic) and the focus of the assessment (operator or technology). These differences will ultimately influence the type of information they are able to provide.

Questionnaires can be used to gauge control desk engineers' (CDEs) reactions, opinions and attitudes to their alarm system, both in general and specific instances. For example to:

- elicit the CDEs' definition of the term 'alarm';
- examine the CDEs' alarm handling activities;
- get information on problems with the alarm system.

Observations provide information about a particular period of activity. Types of information could include:

- how alarm handling fits in with other duties;
- data on the quantity of alarm signals;
- the way in which the information is used.

Scenarios offer a means of examining the potential interaction of the operators, prior to the commissioning of the alarm system. The assessment may consider:

- generic activities;
- operators' responses;
- use of procedures.

The resultant information can be used to make changes if the assessment is conducted early enough in the design programme.

Static assessments provide a means of scrutinizing the alarm system along such dimensions as:

- functional grouping;
- justification;
- location identification;
- physical characteristics;
- response procedures;
- visual presentation;
- alarm setpoints.

Information gained from all of these types of assessment can be used to inform designers of future generations of alarm systems. Each of the methods has both advantages as well as potential pitfalls. For example, questionnaires are quick and easy to administer but the respondents will be self selected and could be very limited. Observations collect objective data of the alarm system in action but the results might not generalize beyond that period of observation. Scenarios can provide useful data before an alarm system is implemented but the emergent behaviour of the alarm system may be unlike that presented in the scenarios. Similarly, static assessments can provide a very thorough and rigorous evaluation of the alarm system but it may fail to report on the (more important) dynamic aspects of the system.

As indicated previously, the different approach does yield different types of information. Therefore the question is not which is the best approach but which is the best combination of approaches. Obviously this will be driven largely by the purpose for the assessment. It is suggested that it is unwise to rely on one approach and a combination of approaches, in a way that is complementary and supportive, is recommended. Where the methods provide overlap in the data gathered, this may be used to substantiate and support evidence from elsewhere. If contradictions do arise, then this would suggest that further investigations are necessary to explore the differences. In this way a clearer picture of the alarm system may emerge. In summary, different approaches offer the designer of alarm systems a novel means of considering the problem, which is essentially human and task-centred, rather than engineering and process-centred.

Definitions

In the course of this book several definitions of the term 'alarm' have been implied. It appears that the definition depends largely upon the purpose it will serve. Usher (chapter 8) and Stanton (chapter 1) considered various definitions but concluded that none was wholly suitable.

Usher (chapter 8) presents the matrix model of alarm definition. This consists of different levels of aspect matrices:

- the required value matrix;
- the tolerance matrix;
- the measured value matrix;
- the errors matrix;
- the criticality matrix.

A combination of these matrices enables Usher to illustrate how he derives the definition of alarm data. He claims that human factors can contribute to this definition in three ways:

1. determining what the human operations are when interacting with the system;
2. determining the critical elements;
3. deciding how to present the information to the user.

Stanton (chapter 1) develops a systems model of alarms and an adequate definition of the term 'alarm' centred on the communication of information. This definition considers that an alarm is:

> an unexpected change in system state, a means of signalling state changes, a means of attracting attention, a means of arousing the operator and a change in the operator's mental state.

These two approaches are perhaps best contrasted by the stimulus-based and response-based models of alarms presented in chapter 1. The approach taken by Usher may be characterized principally by the former and the approach taken by Stanton may be characterized principally by the latter.

Human supervisory control

This book has considered the design of alarm systems mainly within human supervisory control tasks, and while some of the observations and recommendations may generalize to other situations, this must be done with care. Human supervisory control tasks make very special demands upon the human part of the system. For example, the operator is required to monitor developments within the process and intervene at a point that optimizes process safety and efficiency. The process being monitored may be complex, closely coupled, contain feedback loops and be opaque to interrogation. The data presented are likely to be in their raw form and plentiful. It could also be secondary, i.e. not directly measuring the plant component that the operator wishes to know about, but an effect of it. For example, an operator may wish to know about viscosity of a product but may have to infer this from product flow, temperature and stirring speed.

Typically, operators work in teams under a supervisor. Their duties fall under three main operations; carrying out predetermined tasks, dealing with faults and monitoring the plant. Planned activities might include starting up or shutting down plant, cleaning or maintenance tasks. Fault management activities might be dealt with locally or require intervention by plant engineers. Most of the time operators may be monitoring the plant. Apparent inactivity during this phase belies the assimilation of information as the operator checks and tracks the plant. Under such circumstances the operator is waiting for the plant to go off track, and at such time they will be called to intervene. The system they are monitoring may have up to 800 pages of information and up to 20 000 alarms, so there is plenty of information to monitor. The sheer amount of information makes the alarm system an absolute necessity, as the operators could not monitor all of it even if they wanted to.

These tasks place quite a lot of demand upon the operators. The information they seek is often spatially located, reflecting the physical location of the plant. There are also quite high memory demands associated with collecting the information and remembering what has to be done. Human supervisory control tasks appear to demand concurrent fault management, that is dealing with many faults at the same time, so that they can be continuously monitored and priorities updated, with higher priority faults getting more attention. However, human operators seem to prefer serial fault management, dealing with one fault at a time before turning their attention to the next one. This would appear to be a basic incompatibility between human operators and process control tasks.

Zwaga and Hoonhout (chapter 7) suggest that many of the problems that the operator faces can be attributed to a misconception of how they manage the information. They openly question the notion of management-by-exception, i.e. operators assuming a passive role, subservient to the automatic systems and only intervening when called to do so by the alarm system. They propose that instead operators' behaviour is more appropriately characterized as operation-by-awareness, i.e. operators are actively following the progress of the automatic system and continually seeking evidence that everything is going according to expectation. Marshall and Baker (chapter 11) endorse this latter observation, noting that in the normal course of activities operators in control rooms are actively monitoring the state of the plant.

Woods (chapter 5) points out that management of faults in dynamic environments (such as flightdecks, space systems, anaesthetic management in surgery and process control) makes special demands on the reasoning of human operators. He proposes that there are four basic responses to faults:

- to mitigate the consequences;
- to break the propagation paths;
- to terminate the source;
- to clean up the after-effects.

Woods suggests that human reasoning may be suboptimal and presents a model of the reasoning process to show where errors may occur. Cognitive aids developed on the basis of an understanding of human reasoning could be used to support dynamic fault management. However, Woods cautions that the provision of such systems should proceed with care as they could impede human performance under certain circumstances.

Alarm initiated activities

Stanton (chapter 6) introduces the notion of 'alarm initiated activities' (AIA), i.e. those activities that the operator entered into as a direct result of the onset of the alarm. These were first identified through a content analysis of the questionnaire data and subsequently confirmed in observational studies. The stages of AIA (observe, accept, analyse, investigate, correct and monitor) were presented as a framework for a literature review. The main tenet of the chapter was that each stage of AIA makes particular demands upon the alarm system and some stages may interfere with each other.

In the observe stage the alarm has been detected and brought to the operator's attention. In the accept stage the operator acknowledges the receipt of the alarm. In the analysis stage the operator makes a decision of what to do next, typically: ignore it, monitor it, correct it or investigate it. In the investigation stage the operator seeks further information about the nature and cause of the alarm. In the correct stage the operator makes corrective actions to bring the situation under control. In the monitor stage the operator watches the alarm to make sure that the situation has been recovered.

The requirements of the stages are that:

1. attraction is required in the observation stage;
2. time to identify and acknowledge is required in the acceptance stage;
3. information to classify with related context is required in the analysis stage;
4. underlying causes are required in the investigative stage;
5. appropriate action needs to be afforded in the corrective stage; and
6. operational feedback is required in the monitoring stage.

Characteristics of alarm media

Many chapters consider the characteristics of alarm media. These characteristics should be capitalized upon if the media are to be used appropriately rather than arbitrarily assigned. Baber (chapter 12) indicates that there are essentially three different types of information display:

- tell-tales
- advisories
- warnings

He proposes that the assignment of media to these roles only fails when it is inappropriate. The choice in vehicle systems includes:

- reconfigurable visual displays;
- auditory displays;
- speech displays.

The use of speech displays in vehicles was a notorious failure. Baber points out that this was entirely due to the fact that speech was used primarily to display tell-tale (i.e. relatively trivial) information. Clearly the medium needs to fit the purpose.

Edworthy (chapter 2) illustrates the power of using various qualities of the medium in a way that improves the performance of the alarm. She reports that certain characteristics of auditory warnings could be used to make the alarm sound more, or less, urgent. In order of greatest effect on perceived urgency, Edworthy lists these characteristics as:

- speed of the sound;
- number of repeating units;
- fundamental frequency;
- inharmonicity.

Edworthy suggests that people find it relatively easy to distinguish between levels of urgency, and if the sound is mapped appropriately onto the alarm, then this audible information could help individuals prioritize their interventions. This process can be used to design non-confusing auditory warnings (Meredith and Edworthy, chapter 13) and can help people to discriminate between the equipment the warning is being issued from. Hickling (chapter 10) also underlines the need for clear audible coding from a practical point of view. Meredith and Edworthy (chapter 13) also suggest that benefits are to be reaped from spending time training people in the meaning of warnings. This is often neglected, but could assist in enabling them to discriminate between a relatively large set of warnings (Meredith and Edworthy propose up to 12 warnings).

Characteristics of visual alarm media were also considered within the book. Marshall and Baker (chapter 11) compare the relative merits of traditional alarm technology (e.g. annunciator tiles) with the changes brought with the advent of information technology (e.g. lists of alarm messages on visual display units). They suggest that some of the richness of information has been lost with the modern technology. Annunciator tiles present the control room operator with some spatial reference of the alarm (normally associated with the plant topography) and the ability to recognize familiar patterns of alarms associated with a particular diagnosis. This enables a high level abstraction of the plant state to occur, without requiring the plant operator to read individual messages on the tiles. Alarm lists, however, present little more than a time sequence of the alarms appearance. They also require the operator to read them as individual items, as their spatial reference on the list presents no clue as to their identity. Hollywell and Marshall (chapter 3) also question the utility of presenting alarm information in the form of text messages.

Text messages are undoubtably an easy way for the programmer to present alarm information but this ease of display can lead to a proliferation of messages. As was indicated earlier, under certain conditions a rate of 100 alarms per minute can occur. Hollywell and Marshall found that this far exceeded the individual operators' ability to read and classify the information, which they suggest is approximately 30 alarms per minute (Hollywell and Marshall indicate that the operators preferred alarm handling rate is half this, i.e. 15 alarms per minute). Finally, Hoyes and Stanton (chapter 4) suggest that interventions made by human factors professionals in the domain of human supervisory control tasks will result in long term reduction in accident loss, and a homeostasis effect does not appear to occur. This adds some reassurance to changes made in the design of alarm systems, e.g. reducing the rate of presentation as suggested by Hollywell and Marshall.

Future research

Future research may wish to uncover the potential of 'new' media and information presentation methods, such as: hypermedia, virtual reality, video walls and speech synthesis. Hypermedia has an, as yet, unexploited potential for combining information into any format the operator wishes it to take. This may offer a solution to the information overload normally associated with human supervisory control tasks. Virtual reality offers new display methods, such as allowing the operator to 'look around' normally inhospitable environments. It also offers the possibility of directly controlling plant variables with the use of data gloves. Video walls appear to be reintroducing the back panels into control rooms but they can be far more flexible than hard-wired panels ever were. This flexibility allows the operator to maintain and overview whilst examining a particular aspect of the plant in fine detail. Speech synthesis offers the possibility of exploiting a channel of communication that is normally reserved for interacting with other people. However, introduction of these types of technology needs to proceed with caution. There is a danger that if they are implemented inappropriately they could create more problems than they are intended to solve.

However, before this is undertaken, there is still much basic research needed into how humans use and process alarm information, which this book has begun to uncover. There is yet more work to be done on how traditional alarm media may be effectively combined, and how this information fits in with a 'total' information system. The excitement of human factors research is that there is so much to be done. The main conclusion of the research presented within this book is to highlight the need to consider task demands placed upon the human operators when selecting alarm media. This human-centred approach is central to this book and to human factors generally.

Index